Lecture Notes in Mathematics

1927

Editors:
J.-M. Morel, Cachan
F. Takens, Groningen
B. Teissier, Paris

FONDAZIONE
CIME
ROBERTO CONTI

CENTRO INTERNAZIONALE MATEMATICO ESTIVO
INTERNATIONAL MATHEMATICAL SUMMER CENTER

C.I.M.E. means Centro Internazionale Matematico Estivo, that is, International Mathematical Summer Center. Conceived in the early fifties, it was born in 1954 and made welcome by the world mathematical community where it remains in good health and spirit. Many mathematicians from all over the world have been involved in a way or another in C.I.M.E.'s activities during the past years.

So they already know what the C.I.M.E. is all about. For the benefit of future potential users and co-operators the main purposes and the functioning of the Centre may be summarized as follows: every year, during the summer, Sessions (three or four as a rule) on different themes from pure and applied mathematics are offered by application to mathematicians from all countries. Each session is generally based on three or four main courses (24–30 hours over a period of 6-8 working days) held from specialists of international renown, plus a certain number of seminars.

A C.I.M.E. Session, therefore, is neither a Symposium, nor just a School, but maybe a blend of both. The aim is that of bringing to the attention of younger researchers the origins, later developments, and perspectives of some branch of live mathematics.

The topics of the courses are generally of international resonance and the participation of the courses cover the expertise of different countries and continents. Such combination, gave an excellent opportunity to young participants to be acquainted with the most advance research in the topics of the courses and the possibility of an interchange with the world famous specialists. The full immersion atmosphere of the courses and the daily exchange among participants are a first building brick in the edifice of international collaboration in mathematical research.

C.I.M.E. Director
Pietro ZECCA
Dipartimento di Energetica "S. Stecco"
Università di Firenze
Via S. Marta, 3
50139 Florence
Italy
e-mail: zecca@unifi.it

C.I.M.E. Secretary
Elvira MASCOLO
Dipartimento di Matematica
Università di Firenze
viale G.B. Morgagni 67/A
50134 Florence
Italy
e-mail: mascolo@math.unifi.it

For more information see CIME's homepage: http://www.cime.unifi.it

CIME's activity is supported by:

- Istituto Nazionale di Alta Mathematica "F. Severi"
- Ministero dell'Istruzione, dell'Università e delle Ricerca
- Ministero degli Affari Esteri, Direzione Generale per la Promozione e la Cooperazione, Ufficio V

This CIME course was partially supported by: HyKE a Research Training Network (RTN) financed by the European Union in the 5th Framework Programme "Improving the Human Potential" (1HP). Project Reference: Contract Number: HPRN-CT-2002-00282

Luigi Ambrosio · Luis Caffarelli
Michael G. Crandall · Lawrence C. Evans
Nicola Fusco

Calculus of Variations and Nonlinear Partial Differential Equations

Lectures given at the
C.I.M.E. Summer School
held in Cetraro, Italy
June 27–July 2, 2005

With a historical overview by Elvira Mascolo

Editors: Bernard Dacorogna, Paolo Marcellini

 Springer

FONDAZIONE
CIME
ROBERTO CONTI

Luigi Ambrosio
Scuola Normale Superiore
Piazza dei Cavalieri 7
56126 Pisa, Italy
ambrosio@sns.it

Luis Caffarelli
Department of Mathematics
University of Texas at Austin
1 University Station C1200
Austin, TX 78712-0257, USA
caffarel@math.utexas.edu

Michael G. Crandall
Department of Mathematics
University of California
Santa Barbara, CA 93106, USA
crandall@math.ucsb.edu

Bernard Dacorogna
Section de Mathématiques
Ecole Polytechnique Fédérale
  de Lausanne (EPFL)
Station 8
1015 Lausanne, Switzerland
bernard.dacorogna@epfl.ch

Lawrence C. Evans
Department of Mathematics
University of California
Berkeley, CA 94720-3840, USA
evans@math.berkeley.edu

Nicola Fusco
Dipartimento di Matematica
Università degli Studi di Napoli
Complesso Universitario Monte S. Angelo
Via Cintia
80126 Napoli, Italy
n.fusco@unina.it

Paolo Marcellini
Elvira Mascolo
Dipartimento di Matematica
Università di Firenze
Viale Morgagni 67/A
50134 Firenze, Italy
marcellini@math.unifi.it
mascolo@math.unifi.it

ISBN 978-3-540-75913-3   e-ISBN 978-3-540-75914-0

DOI 10.1007/978-3-540-75914-0

Lecture Notes in Mathematics ISSN print edition: 0075-8434
                              ISSN electronic edition: 1617-9692

Library of Congress Control Number: 2007937407

Mathematics Subject Classification (2000): 35Dxx, 35Fxx, 35Jxx, 35Lxx, 49Jxx

© 2008 Springer-Verlag Berlin Heidelberg

Cover design: *design & production* GmbH, Heidelberg

Printed on acid-free paper

9 8 7 6 5 4 3 2 1

springer.com

Preface

We organized this CIME Course with the aim to bring together a group of top leaders on the fields of *calculus of variations* and *nonlinear partial differential equations*. The list of speakers and the titles of lectures have been the following:

- Luigi Ambrosio, *Transport equation and Cauchy problem for non-smooth vector fields.*
- Luis A. Caffarelli, *Homogenization methods for non divergence equations.*
- Michael Crandall, *The infinity-Laplace equation and elements of the calculus of variations in L-infinity.*
- Gianni Dal Maso, *Rate-independent evolution problems in elasto-plasticity: a variational approach.*
- Lawrence C. Evans, *Weak KAM theory and partial differential equations.*
- Nicola Fusco, *Geometrical aspects of symmetrization.*

In the original list of invited speakers the name of Pierre Louis Lions was also included, but he, at the very last moment, could not participate.

The Course, just looking at the number of participants (more than 140, one of the largest in the history of the CIME courses), was a great success; most of them were young researchers, some others were well known mathematicians, experts in the field. The high level of the Course is clearly proved by the quality of notes that the speakers presented for this Springer Lecture Notes.

We also invited Elvira Mascolo, the CIME scientific secretary, to write in the present book an overview of the history of CIME (which she presented at Cetraro) with special emphasis in calculus of variations and partial differential equations.

Most of the speakers are among the world leaders in the field of *viscosity solutions* of partial differential equations, in particular nonlinear pde's of *implicit type*. Our choice has not been random; in fact we and other mathematicians have recently pointed out a theory of *almost everywhere solutions* of pde's of *implicit type*, which is an approach to solve *nonlinear systems* of pde's. Thus this Course has been an opportunity to bring together experts of viscosity solutions and to see some recent developments in the field.

We briefly describe here the articles presented in this Lecture Notes.

Starting from the lecture by Luigi Ambrosio, where the author studies the well-posedness of the Cauchy problem for the homogeneous conservative continuity equation

$$\frac{d}{dt}\mu_t + D_x \cdot (b\mu_t) = 0, \qquad (t, x) \in I \times \mathbb{R}^d$$

and for the transport equation

$$\frac{d}{dt}w_t + b \cdot \nabla w_t = c_t,$$

where $b(t, x) = b_t(x)$ is a given time-dependent vector field in $\mathbb{R}^d$. The interesting case is when $b_t(\cdot)$ is not necessarily Lipschitz and has, for instance, a Sobolev or BV regularity. Vector fields with this "low" regularity show up, for instance, in several PDE's describing the motion of fluids, and in the theory of conservation laws.

The lecture of Luis Caffarelli gave rise to a joint paper with Luis Silvestre; we quote from their introduction:

"When we look at a differential equation in a very irregular media (composite material, mixed solutions, etc.) from very close, we may see a very complicated problem. However, if we look from far away we may not see the details and the problem may look simpler. The study of this effect in partial differential equations is known as *homogenization*. The effect of the inhomogeneities oscillating at small scales is often not a simple average and may be hard to predict: a geodesic in an irregular medium will try to avoid the bad areas, the roughness of a surface may affect in nontrivial way the shapes of drops laying on it, etc... The purpose of these notes is to discuss three problems in homogenization and their interplay.

In the first problem, we consider the homogenization of a free boundary problem. We study the shape of a drop lying on a rough surface. We discuss in what case the homogenization limit converges to a perfectly round drop. It is taken mostly from the joint work with Antoine Mellet *(see the precise references in the article by Caffarelli and Silvestre in this lecture notes)*. The second problem concerns the construction of plane like solutions to the minimal surface equation in periodic media. This is related to homogenization of minimal surfaces. The details can be found in the joint paper with Rafael de la Llave. The third problem concerns existence of homogenization limits for solutions to fully nonlinear equations in ergodic random media. It is mainly based on the joint paper with Panagiotis Souganidis and Lihe Wang.

We will try to point out the main techniques and the common aspects. The focus has been set to the basic ideas. The main purpose is to make this advanced topics as readable as possible."

Michael Crandall presents in his lecture an outline of the theory of the archetypal L^∞ variational problem in the calculus of variations. Namely, given an open $U \subset \mathbb{R}^n$ and $b \in C(\partial U)$, find $u \in C(\overline{U})$ which agrees with the boundary function b on ∂U and minimizes

$$\mathcal{F}_\infty(u, U) := \||Du|\|_{L^\infty(U)}$$

among all such functions. Here $|Du|$ is the Euclidean length of the gradient Du of u. He is also interested in the "Lipschitz constant" functional as well: if K is any subset of $\mathbb{R}^n$ and $u : K \to \mathbb{R}$, its least Lipschitz constant is denoted by

$$\mathrm{Lip}\,(u, K) := \inf \{ L \in \mathbb{R} : |u(x) - u(y)| \leq L\,|x - y|\,, \forall x, y \in K \}\,.$$

One has $\mathcal{F}_\infty(u, U) = \mathrm{Lip}\,(u, U)$ if U is *convex*, but equality does not hold in general.

The author shows that a function which is absolutely minimizing for Lip is also absolutely minimizing for $\mathcal{F}_\infty$ and conversely. It turns out that the absolutely minimizing functions for Lip and $\mathcal{F}_\infty$ are precisely the viscosity solutions of the famous partial differential equation

$$\Delta_\infty u = \sum_{i,j=1}^n u_{x_i} u_{x_j} u_{x_i x_j} = 0\,.$$

The operator Δ_∞ is called the "∞-Laplacian" and *"viscosity solutions"* of the above equation are said to be $\infty-harmonic$.

In his lecture Lawrence C. Evans introduces some new PDE methods developed over the past 6 years in so-called *"weak KAM theory"*, a subject pioneered by J. Mather and A. Fathi. Succinctly put, the goal of this subject is the employing of dynamical systems, variational and PDE methods to find "integrable structures" within general Hamiltonian dynamics. Main references *(see the precise references in the article by Evans in this lecture notes)* are Fathi's forthcoming book and an article by Evans and Gomes.

Nicola Fusco in his lecture presented in this book considers two model functionals: the *perimeter* of a set E in $\mathbb{R}^n$ and the *Dirichlet integral* of a scalar function u. It is well known that on replacing E or u by its *Steiner symmetral* or its *spherical symmetrization*, respectively, both these quantities decrease. This fact is classical when E is a smooth open set and u is a C^1 function. On approximating a set of finite perimeter with smooth open sets or a Sobolev function by C^1 functions, these inequalities can be extended by lower semicontinuity to the general setting. However, an approximation argument gives no information about the equality case. Thus, if one is interested in understanding when equality occurs, one has to carry on a deeper analysis, based on fine properties of sets of finite perimeter and Sobolev functions. Briefly, this is the subject of Fusco's lecture.

Finally, as an appendix to this CIME Lecture Notes, as we said Elvira Mascolo, the CIME scientific secretary, wrote an interesting overview of the history of CIME having in mind in particular *calculus of variations* and PDES.

We are pleased to express our appreciation to the speakers for their excellent lectures and to the participants for contributing to the success of the Summer School. We had at Cetraro an interesting, rich, nice, friendly atmosphere, created by the speakers, the participants and by the CIME organizers; also for this reason we like to thank the Scientific Committee of CIME, and in particular Pietro Zecca (CIME Director) and Elvira Mascolo (CIME Secretary). We also thank Carla Dionisi, Irene Benedetti and Francesco Mugelli, who took care of the day to day organization with great efficiency.

Bernard Dacorogna *and* Paolo Marcellini

Contents

Geometrical Aspects of Symmetrization

CIME Courses on Partial Differential Equations and Calculus of Variations

Transport Equation and Cauchy Problem for Non-Smooth Vector Fields

Luigi Ambrosio

Scuola Normale Superiore
Piazza dei Cavalieri 7, 56126 Pisa, Italy
l.ambrosio@sns.it

1 Introduction

In these lectures we study the well-posedness of the Cauchy problem for the homogeneous conservative continuity equation

(PDE) $$\frac{d}{dt}\mu_t + D_x \cdot (\boldsymbol{b}\mu_t) = 0 \qquad (t,x) \in I \times \mathbb{R}^d$$

and for the transport equation

$$\frac{d}{dt}w_t + \boldsymbol{b} \cdot \nabla w_t = c_t.$$

Here $\boldsymbol{b}(t,x) = \boldsymbol{b}_t(x)$ is a given time-dependent vector field in $\mathbb{R}^d$: we are interested to the case when $\boldsymbol{b}_t(\cdot)$ is not necessarily Lipschitz and has, for instance, a Sobolev or BV regularity. Vector fields with this "low" regularity show up, for instance, in several PDE's describing the motion of fluids, and in the theory of conservation laws.

We are also particularly interested to the well posedness of the system of ordinary differential equations

(ODE) $$\begin{cases} \dot{\gamma}(t) = \boldsymbol{b}_t(\gamma(t)) \\ \gamma(0) = x. \end{cases}$$

In some situations one might hope for a "generic" uniqueness of the solutions of ODE, i.e. for "almost every" initial datum x. An even weaker requirement is the research of a "selection principle", i.e. a strategy to select for $\mathcal{L}^d$-almost every x a solution $\boldsymbol{X}(\cdot,x)$ in such a way that this selection is stable w.r.t. smooth approximations of $\boldsymbol{b}$.

In other words, we would like to know that, whenever we approximate $\boldsymbol{b}$ by smooth vector fields $\boldsymbol{b}^h$, the classical trajectories $\boldsymbol{X}^h$ associated to $\boldsymbol{b}^h$ satisfy

$$\lim_{h \to \infty} \boldsymbol{X}^h(\cdot,x) = \boldsymbol{X}(\cdot,x) \qquad \text{in } C([0,T];\mathbb{R}^d), \text{ for } \mathcal{L}^d\text{-a.e. } x.$$

The following simple example provides an illustration of the kind of phenomena that can occur.

Example 1.1. Let us consider the autonomous ODE

$$\begin{cases} \dot{\gamma}(t) = \sqrt{|\gamma(t)|} \\ \gamma(0) = x_0. \end{cases}$$

Then, solutions of the ODE are not unique for $x_0 = -c^2 < 0$. Indeed, they reach the origin in time $2c$, where can stay for an arbitrary time T, then continuing as $x(t) = \frac{1}{4}(t - T - 2c)^2$. Let us consider for instance the Lipschitz approximation (that could easily be made smooth) of $b(\gamma) = \sqrt{|\gamma|}$ by

$$b_\varepsilon(\gamma) := \begin{cases} \sqrt{|\gamma|} & \text{if } -\infty < \gamma \leq -\varepsilon^2; \\ \varepsilon & \text{if } -\varepsilon^2 \leq \gamma \leq \lambda_\varepsilon - \varepsilon^2 \\ \sqrt{\gamma - \lambda_\varepsilon + 2\varepsilon^2} & \text{if } \lambda_\varepsilon - \varepsilon^2 \leq \gamma < +\infty, \end{cases}$$

with $\lambda_\varepsilon - \varepsilon^2 > 0$. Then, solutions of the approximating ODE's starting from $-c^2$ reach the value $-\varepsilon^2$ in time $t_\varepsilon = 2(c - \varepsilon)$ and then they continue with constant speed ε until they reach $\lambda_\varepsilon - \varepsilon^2$, in time $T_\varepsilon = \lambda_\varepsilon/\varepsilon$. Then, they continue as $\lambda_\varepsilon - 2\varepsilon^2 + \frac{1}{4}(t - t_\varepsilon - T_\varepsilon)^2$.

Choosing $\lambda_\varepsilon = \varepsilon T$, with $T > 0$, by this approximation we select the solutions that don't move, when at the origin, exactly for a time T.

Other approximations, as for instance $b_\varepsilon(\gamma) = \sqrt{\varepsilon + |\gamma|}$, select the solutions that move immediately away from the singularity at $\gamma = 0$. Among all possibilities, this family of solutions $x(t, x_0)$ is singled out by the property that $x(t, \cdot)_\# \mathcal{L}^1$ is absolutely continuous with respect to $\mathcal{L}^1$, so no concentration of trajectories occurs at the origin. To see this fact, notice that we can integrate in time the identity

$$0 = x(t, \cdot)_\# \mathcal{L}^1(\{0\}) = \mathcal{L}^1(\{x_0 : x(t, x_0) = 0\})$$

and use Fubini's theorem to obtain

$$0 = \int \mathcal{L}^1(\{t : x(t, x_0) = 0\}) \, dx_0.$$

Hence, for $\mathcal{L}^1$-a.e. x_0, $x(\cdot, x_0)$ does not stay at 0 for a strictly positive set of times.

We will see that there is a close link between (PDE) and (ODE), first investigated in a nonsmooth setting by Di Perna and Lions in [53].

Let us now make some basic technical remarks on the continuity equation and the transport equation:

Remark 1.1 (Regularity in space of b_t and μ_t). (1) Since the continuity equation (PDE) is in divergence form, it makes sense without *any* regularity requirement on b_t and/or μ_t, provided

$$\int_I \int_A |\boldsymbol{b}_t|\, d|\mu_t|\, dt < +\infty \qquad \forall A \subset\subset \mathbb{R}^d. \tag{1.1}$$

However, when we consider possibly singular measures μ_t, we must take care of the fact that the product $\boldsymbol{b}_t \mu_t$ is sensitive to modifications of $\boldsymbol{b}_t$ in $\mathcal{L}^d$-negligible sets. In the Sobolev or BV case we will consider only measures $\mu_t = w_t \mathcal{L}^d$, so everything is well posed.

(2) On the other hand, due to the fact that the distribution $\boldsymbol{b}_t \cdot \nabla w$ is defined by

$$\langle \boldsymbol{b}_t \cdot \nabla w, \varphi \rangle := -\int_I \int w \langle \boldsymbol{b}_t, \nabla \varphi \rangle dx dt - \int_I \langle D_x \cdot \boldsymbol{b}_t, w_t \varphi_t \rangle dt \quad \varphi \in C_c^\infty(I \times \mathbb{R}^d)$$

(a definition consistent with the case when w_t is smooth) the transport equation makes sense *only* if we assume that $D_x \cdot \boldsymbol{b}_t = \operatorname{div} \boldsymbol{b}_t \mathcal{L}^d$ for $\mathcal{L}^1$-a.e. $t \in I$. See also [28], [31] for recent results on the transport equation when $\boldsymbol{b}$ satisfies a one-sided Lipschitz condition.

Next, we consider the problem of the time continuity of $t \mapsto \mu_t$ and $t \mapsto w_t$.

Remark 1.2 (Regularity in time of μ_t). For any test function $\varphi \in C_c^\infty(\mathbb{R}^d)$, condition (7.11) gives

$$\frac{d}{dt} \int_{\mathbb{R}^d} \varphi\, d\mu_t = \int_{\mathbb{R}^d} \boldsymbol{b}_t \cdot \nabla \varphi\, d\mu_t \in L^1(I)$$

and therefore the map $t \mapsto \langle \mu_t, \varphi \rangle$, for given φ, has a unique uniformly continuous representative in I. By a simple density argument we can find a unique representative $\tilde{\mu}_t$ independent of φ, such that $t \mapsto \langle \tilde{\mu}_t, \varphi \rangle$ is uniformly continuous in I for any $\varphi \in C_c^\infty(\mathbb{R}^d)$. We will always work with this representative, so that μ_t will be well defined *for all t* and even at the endpoints of I.

An analogous remark applies for solutions of the transport equation.

There are some other important links between the two equations:

(1) The transport equation reduces to the continuity equation in the case when $c_t = -w_t \operatorname{div} \boldsymbol{b}_t$.

(2) Formally, one can estabilish a duality between the two equations via the (formal) identity

$$\frac{d}{dt}\int w_t\, d\mu_t = \int \frac{d}{dt} w_t\, d\mu_t + \int \frac{d}{dt}\mu_t w_t$$

$$= \int (-\boldsymbol{b}_t \cdot \nabla w_t + c)\, d\mu_t + \int \boldsymbol{b}_t \cdot \nabla w_t\, d\mu_t = \int c\, d\mu_t.$$

This duality method is a classical tool to prove uniqueness in a sufficiently smooth setting (but see also [28], [31]).

(3) Finally, if we denote by $\boldsymbol{Y}(t, s, x)$ the solution of the ODE at time t, starting from x at the initial times s, *i.e.*

$$\frac{d}{dt}\boldsymbol{Y}(t,s,x) = \boldsymbol{b}_t(\boldsymbol{Y}(t,s,x)), \qquad \boldsymbol{Y}(s,s,x) = x,$$

then $\boldsymbol{Y}(t,\cdot,\cdot)$ are themselves solutions of the transport equation: to see this, it suffices to differentiate the semigroup identity

$$\boldsymbol{Y}(t,s,\boldsymbol{Y}(s,l,x)) = \boldsymbol{Y}(t,l,x)$$

w.r.t. s to obtain, after the change of variables $y = \boldsymbol{Y}(s,l,x)$, the equation

$$\frac{d}{ds}\boldsymbol{Y}(t,s,y) + \boldsymbol{b}_s(y)\cdot\nabla\boldsymbol{Y}(t,s,y) = 0.$$

This property is used in a essential way in [53] to characterize the flow $\boldsymbol{Y}$ and to prove its stability properties. The approach developed here, based on [7], is based on a careful analysis of the measures transported by the flow, and ultimately on the homogeneous continuity equation only.

Acknowledgement. I wish to thank Gianluca Crippa and Alessio Figalli for their careful reading of a preliminary version of this manuscript.

2 Transport Equation and Continuity Equation within the Cauchy-Lipschitz Framework

In this section we recall the classical representation formulas for solutions of the continuity or transport equation in the case when

$$\boldsymbol{b} \in L^1\left([0,T]; W^{1,\infty}(\mathbb{R}^d;\mathbb{R}^d)\right).$$

Under this assumption it is well known that solutions $\boldsymbol{X}(t,\cdot)$ of the ODE are unique and stable. A quantitative information can be obtained by differentiation:

$$\frac{d}{dt}|\boldsymbol{X}(t,x) - \boldsymbol{X}(t,y)|^2 = 2\langle\boldsymbol{b}_t(\boldsymbol{X}(t,x)) - \boldsymbol{b}_t(\boldsymbol{X}(t,y)), \boldsymbol{X}(t,x) - \boldsymbol{X}(t,y)\rangle$$

$$\leq 2\mathrm{Lip}\,(\boldsymbol{b}_t)|\boldsymbol{X}(t,x) - \boldsymbol{X}(t,y)|^2$$

(here $\mathrm{Lip}\,(f)$ denotes the least Lipschitz constant of f), so that Gronwall lemma immediately gives

$$\mathrm{Lip}\,(\boldsymbol{X}(t,\cdot)) \leq \exp\left(\int_0^t \mathrm{Lip}\,(\boldsymbol{b}_s)\,ds\right). \tag{2.1}$$

Turning to the continuity equation, uniqueness of measure-valued solutions can be proved by the duality method. Or, following the techniques developed in these lectures, it can be proved in a more general setting for positive measure-valued solutions (via the superposition principle) and for signed solutions $\mu_t = w_t\mathcal{L}^d$ (via the theory of renormalized solutions). So in this section we focus only on the existence and the representation issues.

The representation formula is indeed very simple:

Proposition 2.1. *For any initial datum $\bar{\mu}$ the solution of the continuity equation is given by*

$$\mu_t := \boldsymbol{X}(t, \cdot)_{\#}\bar{\mu}, \quad i.e. \quad \int_{\mathbb{R}^d} \varphi \, d\mu_t = \int_{\mathbb{R}^d} \varphi(\boldsymbol{X}(t, x)) \, d\bar{\mu}(x). \qquad (2.2)$$

Proof. Notice first that we need only to check the distributional identity $\frac{d}{dt}\mu_t + D_x \cdot (\boldsymbol{b}_t \mu_t) = 0$ on test functions of the form $\psi(t)\varphi(x)$, so that

$$\int_{\mathbb{R}} \psi'(t)\langle \mu_t, \varphi \rangle \, dt + \int_{\mathbb{R}} \psi(t) \int_{\mathbb{R}^d} \langle \boldsymbol{b}_t, \nabla\varphi \rangle \, d\mu_t \, dt = 0.$$

This means that we have to check that $t \mapsto \langle \mu_t, \varphi \rangle$ belongs to $W^{1,1}(0, T)$ for any $\varphi \in C_c^\infty(\mathbb{R}^d)$ and that its distributional derivative is $\int_{\mathbb{R}^d} \langle \boldsymbol{b}_t, \nabla\varphi \rangle \, d\mu_t$. We show first that this map is absolutely continuous, and in particular $W^{1,1}(0, T)$; then one needs only to compute the pointwise derivative. For every choice of finitely many, say n, pairwise disjoint intervals $(a_i, b_i) \subset [0, T]$ we have

$$\sum_{i=1}^{n} |\varphi(\boldsymbol{X}(b_i, x)) - \varphi(\boldsymbol{X}(a_i, x))| \leq \|\nabla\varphi\|_\infty \int_{\cup_i(a_i, b_i)} |\dot{\boldsymbol{X}}(t, x)| \, dt$$

$$\leq \|\nabla\varphi\|_\infty \int_{\cup_i(a_i, b_i)} \sup |\boldsymbol{b}_t| \, dt$$

and therefore an integration with respect to $\bar{\mu}$ gives

$$\sum_{i=1}^{n} |\langle \mu_{b_i} - \mu_{a_i}, \varphi \rangle| \leq \|\nabla\varphi\|_\infty \int_{\cup_i(a_i, b_i)} \sup |\boldsymbol{b}_t| \, dt.$$

The absolute continuity of the integral shows that the right hand side can be made small when $\sum_i(b_i - a_i)$ is small. This proves the absolute continuity. For any x the identity $\dot{\boldsymbol{X}}(t, x) = \boldsymbol{b}_t(\boldsymbol{X}(t, x))$ is fulfilled for $\mathcal{L}^1$-a.e. $t \in [0, T]$. Then, by Fubini's theorem, we know also that for $\mathcal{L}^1$-a.e. $t \in [0, T]$ the previous identity holds for $\bar{\mu}$-a.e. x, and therefore

$$\frac{d}{dt}\langle \mu_t, \varphi \rangle = \frac{d}{dt} \int_{\mathbb{R}^d} \varphi(\boldsymbol{X}(t, x)) \, d\bar{\mu}(x)$$

$$= \int_{\mathbb{R}^d} \langle \nabla\varphi(\boldsymbol{X}(t, x)), \boldsymbol{b}_t(\boldsymbol{X}(t, x)) \rangle \, d\bar{\mu}(x)$$

$$= \langle \boldsymbol{b}_t \mu_t, \nabla\varphi \rangle$$

for $\mathcal{L}^1$-a.e. $t \in [0, T]$.

In the case when $\bar{\mu} = \rho\mathcal{L}^d$ we can say something more, proving that the measures $\mu_t = \boldsymbol{X}(t, \cdot)_{\#}\bar{\mu}$ are absolutely continuous w.r.t. $\mathcal{L}^d$ and computing *explicitly* their density. Let us start by recalling the classical *area formula*: if $f : \mathbb{R}^d \to \mathbb{R}^d$ is a (locally) Lipschitz map, then

$$\int_A g |Jf| \, dx = \int_{\mathbb{R}^d} \sum_{x \in A \cap f^{-1}(y)} g(x) \, dy$$

for any Borel set $A \subset \mathbb{R}^d$, where $Jf = \det \nabla f$ (recall that, by Rademacher theorem, Lipschitz functions are differentiable $\mathcal{L}^d$-a.e.). Assuming in addition that f is 1-1 and onto and that $|Jf| > 0$ $\mathcal{L}^d$-a.e. on A we can set $A = f^{-1}(B)$ and $g = \rho / |Jf|$ to obtain

$$\int_{f^{-1}(B)} \rho \, dx = \int_B \frac{\rho}{|Jf|} \circ f^{-1} \, dy.$$

In other words, we have got a formula for the push-forward:

$$f_\# (\rho \mathcal{L}^d) = \frac{\rho}{|Jf|} \circ f^{-1} \mathcal{L}^d. \qquad (2.3)$$

In our case $f(x) = \boldsymbol{X}(t, x)$ is surely 1-1, onto and Lipschitz. It remains to show that $|J\boldsymbol{X}(t, \cdot)|$ does not vanish: in fact, one can show that $J\boldsymbol{X} > 0$ and

$$\exp \left[-\int_0^t \|[\operatorname{div} \boldsymbol{b}_s]^-\|_\infty \, ds \right] \le J\boldsymbol{X}(t, x) \le \exp \left[\int_0^t \|[\operatorname{div} \boldsymbol{b}_s]^+\|_\infty \, ds \right] \qquad (2.4)$$

for $\mathcal{L}^d$-a.e. x, thanks to the following fact, whose proof is left as an exercise.

Exercise 2.1. If $\boldsymbol{b}$ is smooth, we have

$$\frac{d}{dt} J\boldsymbol{X}(t, x) = \operatorname{div} \boldsymbol{b}_t(\boldsymbol{X}(t, x)) J\boldsymbol{X}(t, x).$$

Hint: use the ODE $\frac{d}{dt} \nabla \boldsymbol{X} = \nabla \boldsymbol{b}_t(\boldsymbol{X}) \nabla \boldsymbol{X}$.

The previous exercise gives that, in the smooth case, $J\boldsymbol{X}(\cdot, x)$ solves a linear ODE with the initial condition $J\boldsymbol{X}(0, x) = 1$, whence the estimates on $J\boldsymbol{X}$ follow. In the general case the upper estimate on $J\boldsymbol{X}$ still holds by a smoothing argument, thanks to the lower semicontinuity of

$$\Phi(v) := \begin{cases} \|Jv\|_\infty & \text{if } Jv \ge 0 \ \mathcal{L}^d\text{-a.e.} \\ +\infty & \text{otherwise} \end{cases}$$

with respect to the w^*-topology of $W^{1,\infty}(\mathbb{R}^d; \mathbb{R}^d)$. This is indeed the supremum of the family of $\Phi_p^{1/p}$, where Φ_p are the *polyconvex* (and therefore lower semicontinuous) functionals

$$\Phi_p(v) := \int_{B_p} |\chi(Jv)|^p \, dx.$$

Here $\chi(t)$, equal to ∞ on $(-\infty, 0)$ and equal to t on $[0, +\infty)$, is l.s.c. and convex. The lower estimate can be obtained by applying the upper one in a time reversed situation.

Now we turn to the representation of solutions of the transport equation:

Proposition 2.2. *If $w \in L^1_{\text{loc}}([0, T] \times \mathbb{R}^d)$ solves*

$$\frac{d}{dt} w_t + b \cdot \nabla w = c \in L^1_{\text{loc}}([0, T] \times \mathbb{R}^d)$$

then, for $\mathcal{L}^d$-a.e. x, we have

$$w_t(X(t, x)) = w_0(x) + \int_0^t c_s(X(s, x))\,ds \qquad \forall t \in [0, T].$$

The (formal) proof is based on the simple observation that

$$\frac{d}{dt} w_t \circ X(t, x) = \frac{d}{dt} w_t(X(t, x)) + \frac{d}{dt} X(t, x) \cdot \nabla w_t(X(t, x))$$

$$= \frac{d}{dt} w_t(X(t, x)) + b_t(X(t, x)) \cdot \nabla w_t(X(t, x))$$

$$= c_t(X(t, x)).$$

In particular, as $X(t, x) = Y(t, 0, x) = [Y(0, t, \cdot)]^{-1}(x)$, we get

$$w_t(y) = w_0(Y(0, t, y)) + \int_0^t c_s(Y(s, t, y))\,ds.$$

We conclude this presentation of the classical theory pointing out two simple local variants of the assumption $b \in L^1([0, T]; W^{1,\infty}(\mathbb{R}^d; \mathbb{R}^d))$ made throughout this section.

Remark 2.1 (First local variant). The theory outlined above still works under the assumptions

$$b \in L^1\left([0, T]; W^{1,\infty}_{\text{loc}}(\mathbb{R}^d; \mathbb{R}^d)\right), \qquad \frac{|b|}{1 + |x|} \in L^1\left([0, T]; L^\infty(\mathbb{R}^d)\right).$$

Indeed, due to the growth condition on b, we still have pointwise uniqueness of the ODE and a uniform local control on the growth of $|X(t, x)|$, therefore we need only to consider a *local* Lipschitz condition w.r.t. x, integrable w.r.t. t.

The next variant will be used in the proof of the superposition principle.

Remark 2.2 (Second local variant). Still keeping the $L^1(W^{1,\infty}_{\text{loc}})$ assumption, and assuming $\mu_t \geq 0$, the second growth condition on $|b|$ can be replaced by a global, but more intrinsic, condition:

$$\int_0^T \int_{\mathbb{R}^d} \frac{|b_t|}{1+|x|} \, d\mu_t \, dt < +\infty. \qquad (2.5)$$

Under this assumption one can show that for $\bar{\mu}$-a.e. x the *maximal* solution $X(\cdot, x)$ of the ODE starting from x is defined up to $t = T$ and still the representation $\mu_t = X(t, \cdot)_{\#}\bar{\mu}$ holds for $t \in [0, T]$.

3 ODE Uniqueness versus PDE Uniqueness

In this section we illustrate some quite general principles, whose application may depend on specific assumptions on b, relating the uniqueness of the ODE to the uniqueness of the PDE. The viewpoint adopted in this section is very close in spirit to Young's theory [85] of generalized surfaces and controls (a theory with remarkable applications also non-linear PDE's [52, 78] and Calculus of Variations [19]) and has also some connection with Brenier's weak solutions of incompressible Euler equations [24], with Kantorovich's viewpoint in the theory of optimal transportation [57, 76] and with Mather's theory [71, 72, 18]: in order to study existence, uniqueness and stability with respect to perturbations of the data of solutions to the ODE, we consider suitable measures in the space of continuous maps, allowing for superposition of trajectories. Then, in some special situations we are able to show that this superposition actually does not occur, but still this "probabilistic" interpretation is very useful to understand the underlying techniques and to give an intrinsic characterization of the flow.

The first very general criterion is the following.

Theorem 3.1. *Let $A \subset \mathbb{R}^d$ be a Borel set. The following two properties are equivalent:*

(a) Solutions of the ODE are unique for any $x \in A$.
(b) Nonnegative measure-valued solutions of the PDE are unique for any $\bar{\mu}$ concentrated in A, i.e. such that $\bar{\mu}(\mathbb{R}^d \setminus A) = 0$.

Proof. It is clear that (b) implies (a), just choosing $\bar{\mu} = \delta_x$ and noticing that two different solutions $X(t)$, $\tilde{X}(t)$ of the ODE induce two different solutions of the PDE, namely $\delta_{X(t)}$ and $\delta_{\tilde{X}(t)}$.

The converse implication is less obvious and requires the superposition principle that we are going to describe below, and that provides the representation

$$\int_{\mathbb{R}^d} \varphi \, d\mu_t = \int_{\mathbb{R}^d} \left(\int_{\Gamma_T} \varphi(\gamma(t)) \, d\eta_x(\gamma) \right) d\mu_0(x),$$

with η_x probability measures concentrated on the absolutely continuous integral solutions of the ODE starting from x. Therefore, when these are unique, the measures η_x are unique (and are Dirac masses), so that the solutions of the PDE are unique.

We will use the shorter notation Γ_T for the space $C\left([0,T];\mathbb{R}^d\right)$ and denote by $e_t : \Gamma_T \to \mathbb{R}^d$ the evaluation maps $\gamma \mapsto \gamma(t)$, $t \in [0,T]$.

Definition 3.1 (Superposition Solutions). *Let $\eta \in \mathcal{M}_+(\mathbb{R}^d \times \Gamma_T)$ be a measure concentrated on the set of pairs (x,γ) such that γ is an absolutely continuous integral solution of the ODE with $\gamma(0) = x$. We define*

$$\langle \mu_t^\eta, \varphi \rangle := \int_{\mathbb{R}^d \times \Gamma_T} \varphi(e_t(\gamma))\, d\eta(x,\gamma) \qquad \forall \varphi \in C_b(\mathbb{R}^d).$$

By a standard approximation argument the identity defining μ_t^η holds for any Borel function φ such that $\gamma \mapsto \varphi(e_t(\gamma))$ is η-integrable (or equivalently any μ_t^η-integrable function φ).

Under the (local) integrability condition

$$\int_0^T \int_{\mathbb{R}^d \times \Gamma_T} \chi_{B_R}(e_t)|\boldsymbol{b}_t(e_t)|\, d\eta\, dt < +\infty \qquad \forall R > 0 \tag{3.1}$$

it is not hard to see that μ_t^η solves the PDE with the initial condition $\bar{\mu} := (\pi_{\mathbb{R}^d})_\# \eta$: indeed, let us check first that $t \mapsto \langle \mu_t^\eta, \varphi \rangle$ is absolutely continuous for any $\varphi \in C_c^\infty(\mathbb{R}^d)$. For every choice of finitely many pairwise disjoint intervals $(a_i, b_i) \subset [0,T]$ we have

$$\sum_{i=1}^n |\varphi(\gamma(b_i)) - \varphi(\gamma(a_i))| \leq \mathrm{Lip}\,(\varphi) \int_{\cup_i(a_i,b_i)} \chi_{B_R}(|e_t(\gamma)|)\boldsymbol{b}_t(e_t(\gamma))|\, dt$$

for η-a.e. (x,γ), with R such that $\mathrm{supp}\,\varphi \subset \overline{B}_R$. Therefore an integration with respect to η gives

$$\sum_{i=1}^n |\langle \mu_{b_i}^\eta, \varphi \rangle - \langle \mu_{a_i}^\eta, \varphi \rangle| \leq \mathrm{Lip}\,(\varphi) \int_{\cup_i(a_i,b_i)} \int_{\mathbb{R}^d \times \Gamma_T} \chi_{B_R}(e_t)|\boldsymbol{b}_t(e_t)|\, d\eta\, dt.$$

The absolute continuity of the integral shows that the right hand side can be made small when $\sum_i(b_i - a_i)$ is small. This proves the absolute continuity.

It remains to evaluate the time derivative of $t \mapsto \langle \mu_t^\eta, \varphi \rangle$: we know that for η-a.e. (x,γ) the identity $\dot{\gamma}(t) = \boldsymbol{b}_t(\gamma(t))$ is fulfilled for $\mathcal{L}^1$-a.e. $t \in [0,T]$. Then, by Fubini's theorem, we know also that for $\mathcal{L}^1$-a.e. $t \in [0,T]$ the previous identity holds for η-a.e. (x,γ), and therefore

$$\frac{d}{dt}\langle \mu_t^\eta, \varphi \rangle = \frac{d}{dt} \int_{\mathbb{R}^d \times \Gamma_T} \varphi(e_t(\gamma))\, d\eta$$

$$= \int_{\mathbb{R}^d \times \Gamma_T} \langle \nabla\varphi(e_t(\gamma)), \boldsymbol{b}_t(e_t(\gamma)) \rangle\, d\eta = \langle \boldsymbol{b}_t\mu_t, \nabla\varphi \rangle \quad \mathcal{L}^1\text{-a.e. in } [0,T].$$

Remark 3.1. Actually the formula defining μ_t^η does not contain x, and so it involves only the projection of η on Γ_T. Therefore one could also consider measures $\boldsymbol{\sigma}$ in Γ_T, concentrated on the set of solutions of the ODE (for an arbitrary initial point x). These two viewpoints are basically equivalent: given $\boldsymbol{\eta}$ one can build $\boldsymbol{\sigma}$ just by projection on Γ_T, and given σ one can consider the conditional probability measures $\boldsymbol{\eta}_x$ concentrated on the solutions of the ODE starting from x induced by the random variable $\gamma \mapsto \gamma(0)$ in Γ_T, the law $\bar{\mu}$ (i.e. the push forward) of the same random variable and recover $\boldsymbol{\eta}$ as follows:

$$\int_{\mathbb{R}^d \times \Gamma_T} \varphi(x, \gamma) \, d\boldsymbol{\eta}(x, \gamma) := \int_{\mathbb{R}^d} \left(\int_{\Gamma_T} \varphi(x, \gamma) \, d\boldsymbol{\eta}_x(\gamma) \right) d\bar{\mu}(x). \tag{3.2}$$

Our viewpoint has been chosen just for technical convenience, to avoid the use, wherever this is possible, of the conditional probability theorem.

By restricting $\boldsymbol{\eta}$ to suitable subsets of $\mathbb{R}^d \times \Gamma_T$, several manipulations with superposition solutions of the continuity equation are possible and useful, and these are not immediate to see just at the level of general solutions of the continuity equation. This is why the following result is interesting.

Theorem 3.2 (Superposition Principle). *Let* $\mu_t \in \mathcal{M}_+(\mathbb{R}^d)$ *solve PDE and assume that*

$$\int_0^T \int_{\mathbb{R}^d} \frac{|\boldsymbol{b}|_t(x)}{1 + |x|} \, d\mu_t \, dt < +\infty.$$

Then μ_t *is a superposition solution, i.e. there exists* $\boldsymbol{\eta} \in \mathcal{M}_+(\mathbb{R}^d \times \Gamma_T)$ *such that* $\mu_t = \mu_t^{\boldsymbol{\eta}}$ *for any* $t \in [0, T]$.

In the proof we use the *narrow* convergence of positive measures, i.e. the convergence with respect to the duality with continuous and bounded functions, and the *easy* implication in Prokhorov compactness theorem: any tight and bounded family $\mathcal{F}$ in $\mathcal{M}_+(X)$ is (sequentially) relatively compact w.r.t. the narrow convergence. Remember that tightness means:

for any $\varepsilon > 0$ there exists $K \subset X$ compact s.t. $\mu(X \setminus K) < \varepsilon \ \forall \mu \in \mathcal{F}$.

A necessary and sufficient condition for tightness is the existence of a *coercive* functional $\Psi : X \to [0, \infty]$ such that $\int \Psi \, d\mu \leq 1$ for any $\mu \in \mathcal{F}$.

Proof. **Step 1 (smoothing).** [58] We mollify μ_t w.r.t. the space variable with a kernel ρ having finite first moment M and support equal to the whole of $\mathbb{R}^d$ (a Gaussian, for instance), obtaining smooth and strictly positive functions μ_t^ε. We also choose a function $\psi : \mathbb{R}^d \to [0, +\infty)$ such that $\psi(x) \to +\infty$ as $|x| \to +\infty$ and

$$\int_{\mathbb{R}^d} \psi(x)\mu_0 * \rho_\varepsilon(x) \, dx \leq 1 \qquad \forall \varepsilon \in (0, 1)$$

and a convex nondecreasing function $\Theta : \mathbb{R}^+ \to \mathbb{R}$ having a more than linear growth at infinity such that

$$\int_0^T \int_{\mathbb{R}^d} \frac{\Theta(|b_t|(x))}{1+|x|} \, d\mu_t dt < +\infty$$

(the existence of Θ is ensured by Dunford-Pettis theorem). Defining

$$\mu_t^\varepsilon := \mu_t * \rho_\varepsilon, \qquad b_t^\varepsilon := \frac{(b_t\mu_t) * \rho_\varepsilon}{\mu_t^\varepsilon},$$

it is immediate that

$$\frac{d}{dt}\mu_t^\varepsilon + D_x \cdot (b_t^\varepsilon \mu_t^\varepsilon) = \frac{d}{dt}\mu_t * \rho_\varepsilon + D_x \cdot (b_t\mu_t) * \rho_\varepsilon = 0$$

and that $b^\varepsilon \in L^1\left([0,T]; W^{1,\infty}_{loc}(\mathbb{R}^d; \mathbb{R}^d)\right)$. Therefore Remark 2.2 can be applied and the representation $\mu_t^\varepsilon = X^\varepsilon(t, \cdot)_\# \mu_0^\varepsilon$ still holds. Then, we define

$$\boldsymbol{\eta}^\varepsilon := (x, X^\varepsilon(\cdot, x))_\# \mu_0^\varepsilon,$$

so that

$$\int_{\mathbb{R}^d} \varphi \, d\mu_t^{\eta_\varepsilon} = \int_{\mathbb{R}^d \times \Gamma_T} \varphi(\gamma(t)) \, d\eta^\varepsilon \tag{3.3}$$

$$= \int_{\mathbb{R}^d} \varphi(X^\varepsilon(t, x)) \, d\mu_0^\varepsilon(x) = \int_{\mathbb{R}^d} \varphi \, d\mu_t^\varepsilon.$$

Step 2 (Tightness). We will be using the inequality

$$((1+|x|)c) * \rho_\varepsilon \leq (1+|x|)c * \rho_\varepsilon + \varepsilon c * \tilde{\rho}_\varepsilon \tag{3.4}$$

for c nonnegative measure and $\tilde{\rho}(y) = |y|\rho(y)$, and

$$\Theta(|b_t^\varepsilon(x)|)\mu_t^\varepsilon(x) \leq (\Theta(|b_t|)\mu_t) * \rho_\varepsilon(x). \tag{3.5}$$

The proof of the first one is elementary, while the proof of the second one follows by applying Jensen's inequality with the convex l.s.c. function $(z, t) \mapsto \Theta(|z|/t)t$ (set to $+\infty$ if $t < 0$, or $t = 0$ and $z \neq 0$, and to 0 if $z = t = 0$) and with the measure $\rho_\varepsilon(x - \cdot)\mathcal{L}^d$.

Let us introduce the functional

$$\Psi(x, \gamma) := \psi(x) + \int_0^T \frac{\Theta(|\dot{\gamma}|)}{1+|\gamma|} \, dt,$$

set to $+\infty$ on $\Gamma_T \setminus AC([0, T]; \mathbb{R}^d)$.

Using Ascoli-Arzelá theorem, it is not hard to show that Ψ is coercive (it suffices to show that $\max|\gamma|$ is bounded on the sublevels $\{\Psi \leq t\}$). Since

$$\int_{\mathbb{R}^d \times \Gamma_T} \int_0^T \frac{\Theta(|\dot{\gamma}|)}{1+|\gamma|} \, dt \, d\eta^\varepsilon(x, \gamma) = \int_0^T \int_{\mathbb{R}^d} \frac{\Theta(|b_t^\varepsilon|)}{1+|x|} \, d\mu_t^\varepsilon \, dt$$

$$\overset{(3.4),(3.5)}{\leq} (1+\varepsilon M) \int_0^T \int_{\mathbb{R}^d} \frac{\Theta(|b_t|(x))}{1+|x|} \, d\mu_t dt$$

and

$$\int_{\mathbb{R}^d \times \Gamma_T} \psi(x) \, d\boldsymbol{\eta}^\varepsilon(x, \gamma) = \int_{\mathbb{R}^d} \psi(x) \, d\mu_0^\varepsilon \le 1$$

we obtain that $\int \Psi \, d\boldsymbol{\eta}^\varepsilon$ is uniformly bounded for $\varepsilon \in (0, 1)$, and therefore Prokhorov compactness theorem tells us that the family $\boldsymbol{\eta}^\varepsilon$ is narrowly sequentially relatively compact as $\varepsilon \downarrow 0$. If $\boldsymbol{\eta}$ is any limit point we can pass to the limit in (3.3) to obtain that $\mu_t = \mu_t^{\boldsymbol{\eta}}$.

Step 3 ($\boldsymbol{\eta}$ is Concentrated on Solutions of the ODE). It suffices to show that

$$\int_{\mathbb{R}^d \times \Gamma_T} \frac{\left| \gamma(t) - x - \int_0^t \boldsymbol{b}_s(\gamma(s)) \, ds \right|}{1 + \max\limits_{[0,T]} |\gamma|} \, d\boldsymbol{\eta} = 0 \tag{3.6}$$

for any $t \in [0, T]$. The technical difficulty is that this test function, due to the lack of regularity of $\boldsymbol{b}$, is not continuous. To this aim, we prove first that

$$\int_{\mathbb{R}^d \times \Gamma_T} \frac{\left| \gamma(t) - x - \int_0^t \boldsymbol{c}_s(\gamma(s)) \, ds \right|}{1 + \max\limits_{[0,T]} |\gamma|} \, d\boldsymbol{\eta} \le \int_0^T \int_{\mathbb{R}^d} \frac{|\boldsymbol{b}_s - \boldsymbol{c}_s|}{1 + |x|} \, d\mu_s ds \tag{3.7}$$

for any continuous function $\boldsymbol{c}$ with compact support. Then, choosing a sequence $(\boldsymbol{c}^n)$ converging to $\boldsymbol{b}$ in $L^1(\nu; \mathbb{R}^d)$, with

$$\int \varphi(s, x) \, d\nu(s, x) := \int_0^T \int_{\mathbb{R}^d} \frac{\varphi(s, x)}{1 + |x|} \, d\mu_s(x) \, ds$$

and noticing that

$$\int_{\mathbb{R}^d \times \Gamma_T} \int_0^T \frac{|\boldsymbol{b}_s(\gamma(s)) - \boldsymbol{c}_s^n(\gamma(s))|}{1 + |\gamma(s)|} \, ds d\boldsymbol{\eta} = \int_0^T \int_{\mathbb{R}^d} \frac{|\boldsymbol{b}_s - \boldsymbol{c}_s^n|}{1 + |x|} \, d\mu_s ds \to 0,$$

we can pass to the limit in (3.7) with $\boldsymbol{c} = \boldsymbol{c}^n$ to obtain (3.6).

It remains to show (3.7). This is a limiting argument based on the fact that (3.6) holds for $\boldsymbol{b}^\varepsilon$, $\boldsymbol{\eta}^\varepsilon$:

$$\int_{\mathbb{R}^d \times \Gamma_T} \frac{\left| \gamma(t) - x - \int_0^t \boldsymbol{c}_s(\gamma(s)) \, ds \right|}{1 + \max\limits_{[0,T]} |\gamma|} \, d\boldsymbol{\eta}^\varepsilon$$

$$= \int_{\mathbb{R}^d} \frac{\left| \boldsymbol{X}^\varepsilon(t, x) - x - \int_0^t \boldsymbol{c}_s(\boldsymbol{X}^\varepsilon(s, x)) \, ds \right|}{1 + \max\limits_{[0,T]} |\boldsymbol{X}^\varepsilon(\cdot, x)|} \, d\mu_0^\varepsilon(x)$$

$$= \int_{\mathbb{R}^d} \frac{\left| \int_0^t \boldsymbol{b}_s^\varepsilon(\boldsymbol{X}^\varepsilon(s, x)) - \boldsymbol{c}_s(\boldsymbol{X}^\varepsilon(s, x)) \, ds \right|}{1 + \max\limits_{[0,T]} |\boldsymbol{X}^\varepsilon(\cdot, x)|} \, d\mu_0^\varepsilon(x) \le \int_0^t \int_{\mathbb{R}^d} \frac{|\boldsymbol{b}_s^\varepsilon - \boldsymbol{c}_s|}{1 + |x|} \, d\mu_s^\varepsilon ds$$

$$\leq \int_0^t \int_{\mathbb{R}^d} \frac{|b_s^\varepsilon - c_s^\varepsilon|}{1 + |x|} \, d\mu_s^\varepsilon ds + \int_0^t \int_{\mathbb{R}^d} \frac{|c_s^\varepsilon - c_s|}{1 + |x|} \, d\mu_s^\varepsilon ds$$

$$\leq \int_0^t \int_{\mathbb{R}^d} \frac{|b_s - c_s|}{1 + |x|} \, d\mu_s ds + \int_0^t \int_{\mathbb{R}^d} \frac{|c_s^\varepsilon - c_s|}{1 + |x|} \, d\mu_s^\varepsilon ds.$$

In the last inequalities we added and subtracted $c_t^\varepsilon := (c_t \mu_t) * \rho_\varepsilon / \mu_t^\varepsilon$. Since $c_t^\varepsilon \to c_t$ uniformly as $\varepsilon \downarrow 0$ thanks to the uniform continuity of c, passing to the limit in the chain of inequalities above we obtain (3.7).

The applicability of Theorem 3.1 is strongly limited by the fact that, on one hand, *pointwise* uniqueness properties for the ODE are known only in very special situations, for instance when there is a Lipschitz or a one-sided Lipschitz (or log-Lipschitz, Osgood...) condition on b. On the other hand, also uniqueness for general measure-valued solutions is known only in special situations. It turns out that in many cases uniqueness of the PDE can only be proved in smaller classes $\mathcal{L}$ of solutions, and it is natural to think that this should reflect into a weaker uniqueness condition at the level of the ODE.

We will see indeed that there is uniqueness in the "selection sense". In order to illustrate this concept, in the following we consider a convex class $\mathcal{L}_b$ of measure-valued solutions $\mu_t \in \mathcal{M}_+(\mathbb{R}^d)$ of the continuity equation relative to b, satifying the following monotonicity property:

$$0 \leq \mu_t' \leq \mu_t \in \mathcal{L}_b \quad \Longrightarrow \quad \mu_t' \in \mathcal{L}_b \tag{3.8}$$

whenever μ_t' still solves the continuity equation relative to b, and the integrability condition

$$\int_0^T \int_{\mathbb{R}^d} \frac{|b_t(x)|}{1 + |x|} \, d\mu_t(x) dt < +\infty.$$

The typical application will be with absolutely continuous measures $\mu_t = w_t \mathcal{L}^d$, whose densities satisfy some quantitative and possibly time-depending bound (*e.g.* $L^\infty(L^1) \cap L^\infty(L^\infty)$).

Definition 3.2 ($\mathcal{L}_b$-Lagrangian Flows). *Given the class $\mathcal{L}_b$, we say that $X(t, x)$ is a $\mathcal{L}_b$-Lagrangian flow starting from $\bar{\mu} \in \mathcal{M}_+(\mathbb{R}^d)$ (at time 0) if the following two properties hold:*

(a) $X(\cdot, x)$ is absolutely continuous solution in $[0, T]$ and satisfies

$$X(t, x) = x + \int_0^t b_s(X(s, x)) \, ds \qquad \forall t \in [0, T]$$

for $\bar{\mu}$-a.e. x;
(b) $\mu_t := X(t, \cdot)_\# \bar{\mu} \in \mathcal{L}_b$.

Heuristically $\mathcal{L}_b$-Lagrangian flows can be thought as suitable selections of the solutions of the ODE (possibly non unique), made in such a way to produce a density in $\mathcal{L}_b$, see Example 1.1 for an illustration of this concept.

We will show that the $\mathcal{L}_b$-Lagrangian flow starting from $\bar{\mu}$ is unique, modulo $\bar{\mu}$-negligible sets, whenever a comparison principle for the PDE holds, in the class $\mathcal{L}_b$ (*i.e.* the inequality between two solutions at $t = 0$ is preserved at later times).

Before stating and proving the uniqueness theorem for $\mathcal{L}_b$-Lagrangian flows, we state two elementary but useful results. The first one is a simple exercise:

Exercise 3.1. Let $\sigma \in \mathcal{M}_+(\Gamma_T)$ and let $D \subset [0,T]$ be a dense set. Show that σ is a Dirac mass in Γ_T iff its projections $(e(t))_\# \sigma$, $t \in D$, are Dirac masses in $\mathbb{R}^d$.

The second one is concerned with a family of measures $\boldsymbol{\eta}_x$:

Lemma 3.1. *Let $\boldsymbol{\eta}_x$ be a measurable family of positive finite measures in Γ_T with the following property: for any $t \in [0,T]$ and any pair of disjoint Borel sets E, $E' \subset \mathbb{R}^d$ we have*

$$\boldsymbol{\eta}_x\left(\{\gamma:\ \gamma(t) \in E\}\right)\boldsymbol{\eta}_x\left(\{\gamma:\ \gamma(t) \in E'\}\right) = 0 \quad \bar{\mu}\text{-a.e. in } \mathbb{R}^d. \tag{3.9}$$

Then $\boldsymbol{\eta}_x$ is a Dirac mass for $\bar{\mu}$-a.e. x.

Proof. Taking into account Exercise 3.1, for a fixed $t \in (0,T]$ it suffices to check that the measures $\lambda_x := \gamma(t)_\# \boldsymbol{\eta}_x$ are Dirac masses for $\bar{\mu}$-a.e. x. Then (3.9) gives $\lambda_x(E)\lambda_x(E') = 0$ $\bar{\mu}$-a.e. for any pair of disjoint Borel sets E, $E' \subset \mathbb{R}^d$. Let $\delta > 0$ and let us consider a partition of $\mathbb{R}^d$ in countably many Borel sets R_i having a diameter less then δ. Then, as $\lambda_x(R_i)\lambda_x(R_j) = 0$ μ-a.e. whenever $i \neq j$, we have a corresponding decomposition of $\bar{\mu}$-almost all of $\mathbb{R}^d$ in Borel sets A_i such that supp $\lambda_x \subset \overline{R}_i$ for any $x \in A_i$ (just take $\{\lambda_x(R_i) > 0\}$ and subtract from him all other sets $\{\lambda_x(R_j) > 0\}$, $j \neq i$). Since δ is arbitrary the statement is proved. $\quad\blacksquare$

Theorem 3.3 (Uniqueness of $\mathcal{L}_b$-Lagrangian Flows). *Assume that the PDE fulfils the comparison principle in $\mathcal{L}_b$. Then the $\mathcal{L}_b$-Lagrangian flow starting from $\bar{\mu}$ is unique, i.e. two different selections $\boldsymbol{X}_1(t,x)$ and $\boldsymbol{X}_2(t,x)$ of solutions of the ODE inducing solutions of the the continuity equation in $\mathcal{L}_b$ satisfy*

$$\boldsymbol{X}_1(\cdot,x) = \boldsymbol{X}_2(\cdot,x) \qquad \text{in } \Gamma_T, \text{ for } \bar{\mu}\text{-a.e. } x.$$

Proof. If the statement were false we could produce a measure $\boldsymbol{\eta}$ not concentrated on a graph inducing a solution $\mu_t^{\boldsymbol{\eta}} \in \mathcal{L}_b$ of the PDE. This is not possible, thanks to the next result. The measure $\boldsymbol{\eta}$ can be built as follows:

$$\boldsymbol{\eta} := \frac{1}{2}(\boldsymbol{\eta}^1 + \boldsymbol{\eta}^2) = \frac{1}{2}\left[(x, \boldsymbol{X}_1(\cdot,x))_\# \bar{\mu} + (x, \boldsymbol{X}_2(\cdot,x))_\# \bar{\mu}\right].$$

Since $\mathcal{L}_b$ is convex we still have $\mu_t^{\boldsymbol{\eta}} = \frac{1}{2}(\mu_t^{\boldsymbol{\eta}^1} + \mu_t^{\boldsymbol{\eta}^2}) \in \mathcal{L}_b$.

Remark 3.2. In the same vein, one can also show that

$$X_1(\cdot, x) = X_2(\cdot, x) \qquad \text{in } \Gamma_T \text{ for } \bar{\mu}_1 \wedge \bar{\mu}_2\text{-a.e. } x$$

whenever X_1, X_2 are $\mathcal{L}_b$-Lagrangian flows starting respectively from $\bar{\mu}_1$ and $\bar{\mu}_2$.

We used the following basic result, having some analogy with Kantorovich's and Mather's theories.

Theorem 3.4. *Assume that the PDE fulfils the comparison principle in $\mathcal{L}_b$. Let $\eta \in \mathcal{M}_+(\mathbb{R}^d \times \Gamma_T)$ be concentrated on the pairs (x, γ) with γ absolutely continuous solution of the ODE, and assume that $\mu_t^\eta \in \mathcal{L}_b$. Then η is concentrated on a graph, i.e. there exists a function $x \mapsto X(\cdot, x) \in \Gamma_T$ such that*

$$\eta = \big(x, X(\cdot, x)\big)_{\#} \bar{\mu}, \quad \text{with} \quad \bar{\mu} := (\pi_{\mathbb{R}^d})_{\#} \eta = \mu_0^\eta.$$

Proof. We use the representation (3.2) of η, given by the disintegration theorem, the criterion stated in Lemma 3.1 and argue by contradiction. If the thesis is false then η_x is not a Dirac mass in a set of $\bar{\mu}$ positive measure and we can find $t \in (0, T]$, disjoint Borel sets E, $E' \subset \mathbb{R}^d$ and a Borel set C with $\bar{\mu}(C) > 0$ such that

$$\eta_x \left(\{\gamma : \ \gamma(t) \in E\}\right) \eta_x \left(\{\gamma : \ \gamma(t) \in E'\}\right) > 0 \qquad \forall x \in C.$$

Possibly passing to a smaller set having still strictly positive $\bar{\mu}$ measure we can assume that

$$0 < \eta_x(\{\gamma : \ \gamma(t) \in E\}) \leq M \eta_x(\{\gamma : \ \gamma(t) \in E'\}) \qquad \forall x \in C \qquad (3.10)$$

for some constant M. We define measures η^1, η^2 whose disintegrations η_x^1, η_x^2 are given by

$$\eta_x^1 := \chi_C(x) \eta_x \llcorner \{\gamma : \ \gamma(t) \in E\}, \qquad \eta_x^2 := M \chi_C(x) \eta_x \llcorner \{\gamma : \ \gamma(t) \in E'\}$$

and denote by μ_t^i the (superposition) solutions of the continuity equation induced by η^i. Then

$$\mu_0^1 = \eta_x(\{\gamma : \ \gamma(t) \in E\}) \bar{\mu} \llcorner C, \qquad \mu_0^2 = M \eta_x(\{\gamma : \ \gamma(t) \in E'\}) \bar{\mu} \llcorner C,$$

so that (3.10) yields $\mu_0^1 \leq \mu_0^2$. On the other hand, μ_t^1 is orthogonal to μ_t^2: precisely, denoting by η_{tx} the image of η_x under the map $\gamma \mapsto \gamma(t)$, we have

$$\mu_t^1 = \int_C \eta_{tx} \llcorner E \, d\mu(x) \perp M \int_C \eta_{tx} \llcorner E' \, d\mu(x) = \mu_t^2.$$

Notice also that $\mu_t^i \leq \mu_t$ and so the monotonicity assumption (3.8) on $\mathcal{L}_b$ gives $\mu_t^i \in \mathcal{L}_b$. This contradicts the assumption on the validity of the comparison principle in $\mathcal{L}_b$.

Now we come to the *existence* of $\mathcal{L}_b$-Lagrangian flows.

Theorem 3.5 (Existence of $\mathcal{L}_b$-Lagrangian Flows). *Assume that the PDE fulfils the comparison principle in $\mathcal{L}_b$ and that for some $\bar{\mu} \in \mathcal{M}_+(\mathbb{R}^d)$ there exists a solution $\mu_t \in \mathcal{L}_b$ with $\mu_0 = \bar{\mu}$. Then there exists a (unique) $\mathcal{L}_b$-Lagrangian flow starting from $\bar{\mu}$.*

Proof. By the superposition principle we can represent μ_t as $(e_t)_{\#}\eta$ for some $\eta \in \mathcal{M}_+(\mathbb{R}^d \times \Gamma_T)$ concentrated on pairs (x, γ) solutions of the ODE. Then, Theorem 3.4 tells us that η is concentrated on a graph, i.e. there exists a function $x \mapsto \boldsymbol{X}(\cdot, x) \in \Gamma_T$ such that

$$\left(x, \boldsymbol{X}(\cdot, x)\right)_{\#}\bar{\mu} = \eta.$$

Pushing both sides via e_t we obtain

$$\boldsymbol{X}(t, \cdot)_{\#}\bar{\mu} = (e_t)_{\#}\eta = \mu_t \in \mathcal{L}_b,$$

and therefore $\boldsymbol{X}$ is a $\mathcal{L}_b$-Lagrangian flow.

Finally, let us discuss the *stability* issue. This is particularly relevant, as we will see, in connection with the applications to PDE's.

Definition 3.3 (Convergence of Velocity Fields). We define the convergence of b^h to b in a indirect way, defining rather a convergence of $\mathcal{L}_{b^h}$ to $\mathcal{L}_b$: we require that

$$b^h \mu_t^h \rightharpoonup b\mu_t \text{ in } (0, T) \times \mathbb{R}^d \qquad \text{and} \qquad \mu_t \in \mathcal{L}_b$$

whenever $\mu_t^h \in \mathcal{L}_{b^h}$ and $\mu_t^h \to \mu_t$ narrowly for all $t \in [0, T]$.

For instance, in the typical case when $\mathcal{L}$ is bounded and closed, w.r.t the weak* topology, in $L^\infty(L^1) \cap L^\infty(L^\infty)$, and

$$\mathcal{L}_c := \mathcal{L} \cap \left\{ w : \frac{d}{dt}w + D_x \cdot (cw) = 0 \right\}$$

the implication is fulfilled whenever $b^h \to b$ strongly in L^1_{loc}.

The natural convergence for the stability theorem is *convergence in measure*. Let us recall that a Y-valued sequence (v_h) is said to converge in $\bar{\mu}$-measure to v if

$$\lim_{h \to \infty} \bar{\mu}\left(\{d_Y(v_h, v) > \delta\}\right) = 0 \qquad \forall \delta > 0.$$

This is equivalent to the L^1 convergence to 0 of the $\mathbb{R}^+$-valued maps $1 \wedge d_Y(v_h, v)$.

Recall also that convergence $\bar{\mu}$-a.e. implies convergence in measure, and that the converse implication is true passing to a suitable subsequence.

Theorem 3.6 (Stability of $\mathcal{L}$-Lagrangian Flows). *Assume that*

(i) $\mathcal{L}_{b^h}$ *converge to* $\mathcal{L}_b$;

(ii) $\boldsymbol{X}^h$ *are* $\mathcal{L}_{b^h}$*-flows relative to* $\boldsymbol{b}^h$ *starting from* $\bar{\mu} \in \mathcal{M}_+(\mathbb{R}^d)$ *and* $\boldsymbol{X}$ *is the* $\mathcal{L}_b$*-flow relative to* $\boldsymbol{b}$ *starting from* $\bar{\mu}$;

(iii) *setting* $\mu_t^h := \boldsymbol{X}^h(t, \cdot)_{\#}\bar{\mu}$, *we have*

$$\mu_t^h \to \mu_t \qquad narrowly \ as \ h \to \infty \ for \ all \ t \in [0, T] \qquad (3.11)$$

$$\limsup_{h \to \infty} \int_0^T \int_{\mathbb{R}^d} \frac{\Theta(|\boldsymbol{b}_t^h|)}{1 + |x|} d\mu_t^h dt \leq \int_0^T \int_{\mathbb{R}^d} \frac{\Theta(|\boldsymbol{b}_t|)}{1 + |x|} d\mu_t dt < +\infty \qquad (3.12)$$

for some strictly convex function $\Theta : \mathbb{R}^+ \to \mathbb{R}$ *having a more than linear growth at infinity;*

(iv) *the PDE fulfils the comparison principle in* $\mathcal{L}_b$.

Then $\mu_t = \boldsymbol{X}(t, \cdot)_{\#}\bar{\mu}$ *and* $x \mapsto \boldsymbol{X}^h(\cdot, x)$ *converge to* $x \mapsto \boldsymbol{X}(\cdot, x)$ *in* $\bar{\mu}$*-measure, i.e.*

$$\lim_{h \to \infty} \int_{\mathbb{R}^d} 1 \wedge \sup_{[0,T]} |\boldsymbol{X}^h(\cdot, x) - \boldsymbol{X}(\cdot, x)| \, d\bar{\mu}(x) = 0.$$

Proof. Following the same strategy used in the proof of the superposition principle, we push $\bar{\mu}$ onto the graph of the map $x \mapsto \boldsymbol{X}^h(\cdot, x)$, i.e.

$$\boldsymbol{\eta}^h := \left(x, \boldsymbol{X}^h(\cdot, x)\right)_{\#} \bar{\mu}$$

and we obtain, using (3.12) and the same argument used in Step 2 of the proof of the superposition principle, that $\boldsymbol{\eta}^h$ is tight in $\mathcal{M}_+(\mathbb{R}^d \times \Gamma_T)$.

Let now $\boldsymbol{\eta}$ be any limit point of $\boldsymbol{\eta}^h$. Using the same argument used in Step 3 of the proof of the superposition principle and (3.12) we obtain that η is concentrated on pairs (x, γ) with γ absolutely continuous solution of the ODE relative to $\boldsymbol{b}$ starting from x. Indeed, this argument was using only the property

$$\lim_{h \to \infty} \int_0^T \int_{\mathbb{R}^d} \frac{|\boldsymbol{b}_t^h - \boldsymbol{c}_t|}{1 + |x|} d\mu_t^h \, dt = \int_0^T \int_{\mathbb{R}^d} \frac{|\boldsymbol{b}_t - \boldsymbol{c}_t|}{1 + |x|} d\mu_t \, dt$$

for any continuous function $\boldsymbol{c}$ with compact support in $(0, T) \times \mathbb{R}^d$, and this property is ensured by Lemma 3.3 below.

Let $\mu_t := (e_t)_{\#}\boldsymbol{\eta}$ and notice that $\mu_t^h = (e_t)_{\#}\boldsymbol{\eta}^h$, hence $\mu_t^h \to \mu_t$ narrowly for any $t \in [0, T]$. As $\mu_t^h \in \mathcal{L}_{b^h}$, assumption (i) gives that $\mu_t \in \mathcal{L}_b$ and assumption (iv) together with Theorem 3.4 imply that η is concentrated on the graph of the map $x \mapsto \boldsymbol{X}(\cdot, x)$, where $\boldsymbol{X}$ is the unique $\mathcal{L}_b$-Lagrangian flow. We have thus obtained that

$$\left(x, \boldsymbol{X}^h(\cdot, x)\right)_{\#}\bar{\mu} \quad \rightharpoonup \quad \left(x, \boldsymbol{X}(\cdot, x)\right)_{\#}\bar{\mu}.$$

By applying the following general principle we conclude.

Lemma 3.2 (Narrow Convergence and Convergence in Measure). *Let*

$v_h, v : X \to Y$ *be Borel maps and let* $\bar{\mu} \in \mathcal{M}_+(X)$. *Then* $v_h \to v$ *in* $\bar{\mu}$-*measure iff*

$$(x, v_h(x))_{\#}\bar{\mu} \text{ converges to } (x, v(x))_{\#}\bar{\mu} \text{ narrowly in } \mathcal{M}_+(X \times Y).$$

Proof. If $v_h \to v$ in $\bar{\mu}$-measure then $\varphi(x, v_h(x))$ converges in $L^1(\bar{\mu})$ to $\varphi(x, v(x))$, and we immediately obtain the convergence of the push-forward measures. Conversely, let $\delta > 0$ and, for any $\varepsilon > 0$, let $w \in C_b(X; Y)$ be such that $\bar{\mu}(\{v \neq w\}) \leq \varepsilon$. We define

$$\varphi(x, y) := 1 \wedge \frac{d_Y(y, w(x))}{\delta} \in C_b(X \times Y)$$

and notice that

$$\bar{\mu}(\{v \neq w\}) + \int_{X \times Y} \varphi \, d(x, v_h(x))_{\#}\bar{\mu} \geq \bar{\mu}(\{d_Y(v, v_h) > \delta\}),$$

$$\int_{X \times Y} \varphi \, d(x, v(x))_{\#}\bar{\mu} \leq \bar{\mu}(\{w \neq v\}).$$

Taking into account the narrow convergence of the push-forward we obtain that

$$\limsup_{h \to \infty} \bar{\mu}(\{d_Y(v, v_h) > \delta\}) \leq 2\bar{\mu}(\{w \neq v\}) \leq 2\varepsilon$$

and since ε is arbitrary the proof is achieved.

Lemma 3.3. *Let* $A \subset \mathbb{R}^m$ *be an open set, and let* $\sigma^h \in \mathcal{M}_+(A)$ *be narrowly converging to* $\sigma \in \mathcal{M}_+(A)$. *Let* $\boldsymbol{f}^h \in L^1(A, \sigma^h, \mathbb{R}^k)$, $\boldsymbol{f} \in L^1(A, \sigma, \mathbb{R}^k)$ *and assume that*

(i) $\boldsymbol{f}^h \sigma^h$ *weakly converge, in the duality with* $C_c(A; \mathbb{R}^k)$, *to* $\boldsymbol{f}\sigma$;
(ii) $\limsup_{h \to \infty} \int_A \Theta(|\boldsymbol{f}^h|) \, d\sigma^h \leq \int_A \Theta(|\boldsymbol{f}|) \, d\sigma < +\infty$ *for some strictly convex function* $\Theta : \mathbb{R}^+ \to \mathbb{R}$ *having a more than linear growth at infinity.*

Then $\int_A |\boldsymbol{f}^h - \boldsymbol{c}| \, d\sigma^h \to \int_A |\boldsymbol{f} - \boldsymbol{c}| \, d\sigma$ *for any* $\boldsymbol{c} \in C_b(A; \mathbb{R}^k)$.

Proof. We consider the measures $\nu^h := (x, \boldsymbol{f}^h(x))_{\#}\sigma^h$ in $A \times \mathbb{R}^k$ and we assume, possibly extracting a subsequence, that $\nu^h \to \nu$, with $\nu \in \mathcal{M}_+(A \times \mathbb{R}^k)$, in the duality with $C_c(A \times \mathbb{R}^k)$. Using condition (ii), the narrow convergence of σ^h and a truncation argument it is easy to see that the convergence actually occurs for any continuous test function $\psi(x, y)$ satisfying

$$\lim_{|y| \to \infty} \frac{\sup_x |\psi(x, y)|}{\Theta(|y|)} = 0.$$

Furthermore, for nonnegative continuous functions ψ, we have also

$$\int_{A\times\mathbb{R}^k} \psi \, d\nu \le \liminf_{h\to\infty} \int_{A\times\mathbb{R}^k} \psi \, d\nu_h. \tag{3.13}$$

Then, choosing test functions $\psi = \psi(x) \in C_b(A)$, the convergence of σ^h to σ gives

$$\int_{A\times\mathbb{R}^k} \psi \, d\nu = \int_A \psi \, d\sigma$$

and therefore, according to the disintegration theorem, we can represent ν as

$$\int_{A\times\mathbb{R}^k} \psi(x,y) \, d\nu(x,y) = \int_A \left(\int_{\mathbb{R}^k} \psi(x,y) \, d\nu_x(y) \right) d\sigma(x) \tag{3.14}$$

for a suitable Borel family of probability measures ν_x in $\mathbb{R}^k$. Next, we can use $\psi(x)y_j$ as test functions and assumption (i), to obtain

$$\lim_{h\to\infty} \int_A f_j^h \psi \, d\mu^h = \lim_{h\to\infty} \int_{A\times\mathbb{R}^k} \psi(x)y_j \, d\nu^h = \int_A \psi(x) \left(\int_{\mathbb{R}^k} y_j \, d\nu_x(y) \right) d\sigma(x).$$

As ψ and j are arbitrary, this means that the first moment ν_x, i.e. $\int y \, d\nu_x$, is equal to $f(x)$ for σ-a.e. x.

On the other hand, choosing $\psi(y) = \Theta(|y|)$ as test function in (3.13), assumption (ii) gives

$$\int_A \int_{\mathbb{R}^k} \Theta(|y|) \, d\nu_x(y) \, d\sigma(x) \le \liminf_{h\to\infty} \int_{A\times\mathbb{R}^k} \Theta(|y|) \, d\nu^h =$$
$$= \limsup_{h\to\infty} \int_A \Theta(|f^h|) \, d\sigma^h =$$
$$= \int_A \Theta(|f|) \, d\sigma,$$

hence $\int \Theta(|y|) \, d\nu_x = f(x) = \Theta(|\int y \, d\nu_x|)$ for σ-a.e. x. As Θ is strictly convex, this can happen only if $\nu_x = \delta_{f(x)}$ for σ-a.e. x.

Finally, taking into account the representation (3.14) of ν with $\nu_x = \delta_{f(x)}$, the convergence statement can be achieved just choosing the test function $\psi(x,y) = |y - c(x)|$.

4 Vector Fields with a Sobolev Spatial Regularity

Here we discuss the well-posedness of the continuity or transport equations assuming the $b_t(\cdot)$ has a Sobolev regularity, following [53]. Then, the general theory previously developed provides existence, uniqueness and stability of the $\mathcal{L}$-Lagrangian flow, with $\mathcal{L} := L^\infty(L^1) \cap L^\infty(L^\infty)$. We denote by $I \subset \mathbb{R}$ an open interval.

Definition 4.1 (Renormalized Solutions). *Let $b \in L^1_{\mathrm{loc}} \left(I; L^1_{\mathrm{loc}}(\mathbb{R}^d; \mathbb{R}^d) \right)$ be such that $D \cdot b_t = \mathrm{div}\, b_t \mathcal{L}^d$ for $\mathcal{L}^1$-a.e. $t \in I$, with*

$$\mathrm{div}\, b_t \in L^1_{\mathrm{loc}} \left(I; L^1_{\mathrm{loc}}(\mathbb{R}^d) \right).$$

Let $w \in L^\infty_{\text{loc}}\left(I; L^\infty_{\text{loc}}(\mathbb{R}^d)\right)$ and assume that

$$c := \frac{d}{dt}w + \boldsymbol{b} \cdot \nabla w \in L^1_{\text{loc}}(I \times \mathbb{R}^d). \tag{4.1}$$

Then, we say that w is a renormalized solution of (4.1) if

$$\frac{d}{dt}\beta(w) + \boldsymbol{b} \cdot \nabla\beta(w) = c\beta'(w) \qquad \forall \beta \in C^1(\mathbb{R}).$$

Equivalently, recalling the definition of the distribution $\boldsymbol{b} \cdot \nabla w$, the definition could be given in a conservative form, writing

$$\frac{d}{dt}\beta(w) + D_x \cdot (\boldsymbol{b}\beta(w)) = c\beta'(w) + \operatorname{div} \boldsymbol{b}_t\beta(w).$$

Notice also that the concept makes sense, choosing properly the class of "test" functions β, also for w that do not satisfy (4.1), or are not even locally integrable. This is particularly relevant in connection with DiPerna-Lions's existence theorem for Boltzmann equation , or with the case when w is the characteristic of an unbounded vector field $\boldsymbol{b}$.

This concept is also reminiscent of Kruzhkov's concept of *entropy* solution for a scalar conservation law

$$\frac{d}{dt}u + D_x \cdot (\boldsymbol{f}(u)) = 0 \qquad u : (0, +\infty) \times \mathbb{R}^d \to \mathbb{R}.$$

In this case only a distributional one-sided inequality is required:

$$\frac{d}{dt}\eta(u) + D_x \cdot (\boldsymbol{q}(u)) \leq 0$$

for any convex entropy-entropy flux pair $(\eta, \boldsymbol{q})$ (i.e. η is convex and $\eta'\boldsymbol{f}' = \boldsymbol{q}'$).

Remark 4.1 (Time Continuity). Using the fact that both $t \mapsto w_t$ and $t \mapsto \beta(w_t)$ have a uniformly continuous representative (w.r.t. the $w^* - L^\infty$ topology), we obtain that, for any renormalized solution w, $t \mapsto w_t$ has a unique representative which is continuous w.r.t. the L^1_{loc} topology. The proof follows by a classical weak-strong convergence argument:

$$f_n \rightharpoonup f, \quad \beta(f_n) \rightharpoonup \beta(f) \qquad \Longrightarrow \qquad f_n \to f$$

provided β is *strictly* convex. In the case of scalar conservation laws there are analogous results [82], [73].

Using the concept of renormalized solution we can prove a comparison principle in the following natural class $\mathcal{L}$:

$$\mathcal{L} := \left\{ w \in L^\infty\left([0,T]; L^1(\mathbb{R}^d)\right) \cap L^\infty\left([0,T]; L^\infty(\mathbb{R}^d)\right) : \tag{4.2} \right.$$
$$\left. w \in C\left([0,T]; w^* - L^\infty(\mathbb{R}^d)\right) \right\}.$$

Theorem 4.1 (Comparison Principle). *Assume that*

$$\frac{|\boldsymbol{b}|}{1+|x|} \in L^1\left([0,T]; L^\infty(\mathbb{R}^d)\right) + L^1\left([0,T]; L^1(\mathbb{R}^d)\right), \tag{4.3}$$

that $D \cdot \boldsymbol{b}_t = \operatorname{div} \boldsymbol{b}_t \mathcal{L}^d$ *for* $\mathcal{L}^1$*-a.e.* $t \in [0,T]$, *and that*

$$[\operatorname{div} \boldsymbol{b}_t]^- \in L^1_{\text{loc}}\left([0,T) \times \mathbb{R}^d\right). \tag{4.4}$$

Setting $\boldsymbol{b}_t \equiv 0$ *for* $t < 0$, *assume in addition that* any *solution of* (4.1) *in* $(-\infty, T) \times \mathbb{R}^d$ *is renormalized. Then the comparison principle for the continuity equation holds in the class* $\mathcal{L}$.

Proof. By the linearity of the equation, it suffices to show that $w \in \mathcal{L}$ and $w_0 \leq 0$ implies $w_t \leq 0$ for any $t \in [0,T]$. We extend first the PDE to negative times, setting $w_t = w_0$. Then, fix a cut-off function $\varphi \in C_c^\infty(\mathbb{R}^d)$ with supp $\varphi \subset \overline{B}_2(0)$ and $\varphi \equiv 1$ on $B_1(0)$, and the renormalization functions

$$\beta_\epsilon(t) := \sqrt{\epsilon^2 + (t^+)^2} - \epsilon \in C^1(\mathbb{R}).$$

Notice that

$$\beta_\epsilon(t) \uparrow t^+ \quad \text{as } \epsilon \downarrow 0, \qquad t\beta_\epsilon'(t) - \beta_\epsilon(t) \in [0, \epsilon]. \tag{4.5}$$

We know that

$$\frac{d}{dt}\beta_\epsilon(w_t) + D_x \cdot (\boldsymbol{b}\beta_\epsilon(w_t)) = \operatorname{div} \boldsymbol{b}_t(\beta_\epsilon(w_t) - w_t\beta_\epsilon'(w_t))$$

in the sense of distributions in $(-\infty, T) \times \mathbb{R}^d$. Plugging $\varphi_R(\cdot) := \varphi(\cdot/R)$, with $R \geq 1$, into the PDE we obtain

$$\frac{d}{dt}\int_{\mathbb{R}^d} \varphi_R\beta_\epsilon(w_t)\, dx = \int_{\mathbb{R}^d} \beta_\epsilon(w_t)\langle \boldsymbol{b}_t, \nabla\varphi_R\rangle\, dx + \int_{\mathbb{R}^d} \varphi_R\operatorname{div} \boldsymbol{b}_t(\beta_\epsilon(w_t) - w_t\beta_\epsilon'(w_t))\, dx.$$

Splitting $\boldsymbol{b}$ as $\boldsymbol{b}_1 + \boldsymbol{b}_2$, with

$$\frac{\boldsymbol{b}_1}{1+|x|} \in L^1\left([0,T]; L^\infty(\mathbb{R}^d)\right) \quad \text{and} \quad \frac{\boldsymbol{b}_2}{1+|x|} \in L^1\left([0,T]; L^1(\mathbb{R}^d)\right)$$

and using the inequality

$$\frac{1}{R}\chi_{\{R \leq |x| \leq 2R\}} \leq \frac{3}{1+|x|}\chi_{\{R \leq |x|\}}$$

we can estimate the first integral in the right hand side with

$$3\|\nabla\varphi\|_\infty\|\frac{\boldsymbol{b}_{1t}}{1+|x|}\|_\infty \int_{\{|x| \geq R\}} |w_t|\, dx + 3\|\nabla\varphi\|_\infty\|w_t\|_\infty \int_{\{|x| \geq R\}} \frac{|\boldsymbol{b}_{1t}|}{1+|x|}\, dx.$$

The second integral can be estimated with

$$\varepsilon \int_{\mathbb{R}^d} \varphi_R [\operatorname{div} \boldsymbol{b}_t]^- \, dx,$$

Passing to the limit first as $\varepsilon \downarrow 0$ and then as $R \to +\infty$ and using the integrability assumptions on b and w we get

$$\frac{d}{dt} \int_{\mathbb{R}^d} w_t^+ \, dx \leq 0$$

in the distribution sense in $\mathbb{R}$. Since the function vanishes for negative times, this suffices to conclude using Gronwall lemma.

Remark 4.2. It would be nice to have a completely non-linear comparison principle between renormalized solutions, as in the Kruzkhov theory. Here, on the other hand, we rather used the fact that the difference of the two solutions is renormalized.

In any case, Di Perna and Lions proved that all distributional solutions are renormalized when there is a Sobolev regularity with respect to the spatial variables.

Theorem 4.2. *Let* $b \in L^1_{\mathrm{loc}} \left(I; W^{1,1}_{\mathrm{loc}}(\mathbb{R}^d; \mathbb{R}^d) \right)$ *and let* $w \in L^\infty_{\mathrm{loc}}(I \times \mathbb{R}^d)$ *be a distributional solution of* (4.1). *Then* w *is a renormalized solution.*

Proof. We mollify with respect to the spatial variables and we set

$$r^\varepsilon := (\boldsymbol{b} \cdot \nabla w) * \rho_\varepsilon - \boldsymbol{b} \cdot (\nabla (w * \rho_\varepsilon)), \qquad w^\varepsilon := w * \rho_\varepsilon$$

to obtain

$$\frac{d}{dt} w^\varepsilon + \boldsymbol{b} \cdot \nabla w^\varepsilon = c * \rho_\varepsilon - r^\varepsilon.$$

By the smoothness of w^ε w.r.t. x, the PDE above tells us that $\frac{d}{dt} w_t^\varepsilon \in L^1_{\mathrm{loc}}$, therefore $w^\varepsilon \in W^{1,1}_{\mathrm{loc}}(I \times \mathbb{R}^d)$ and we can apply the standard chain rule in Sobolev spaces, getting

$$\frac{d}{dt} \beta(w^\varepsilon) + \boldsymbol{b} \cdot \nabla \beta(w^\varepsilon) = \beta'(w^\varepsilon) c * \rho_\varepsilon - \beta'(w^\varepsilon) r^\varepsilon.$$

When we let $\varepsilon \downarrow 0$ the convergence in the distribution sense of all terms in the identity above is trivial, with the exception of the last one. To ensure its convergence to zero, it seems necessary to show that $r^\varepsilon \to 0$ strongly in L^1_{loc} (remember that $\beta'(w^\varepsilon)$ is locally equibounded w.r.t. ε). This is indeed the case, and it is exactly here that the Sobolev regularity plays a role.

Proposition 4.1 (Strong convergence of commutators). *If* $w \in L^\infty_{\mathrm{loc}}$ $(I \times \mathbb{R}^d)$ *and* $\boldsymbol{b} \in L^1_{\mathrm{loc}} \left(I; W^{1,1}_{\mathrm{loc}}(\mathbb{R}^d; \mathbb{R}^d) \right)$ *we have*

$$L^1_{\mathrm{loc}}\text{-}\lim_{\varepsilon \downarrow 0} (\boldsymbol{b} \cdot \nabla w) * \rho_\varepsilon - \boldsymbol{b} \cdot (\nabla (w * \rho_\varepsilon)) = 0.$$

Proof. Playing with the definitions of $\boldsymbol{b} \cdot \nabla w$ and convolution product of a distribution and a smooth function, one proves first the identity

$$r^\varepsilon(t,x) = \int_{\mathbb{R}^d} w(t, x - \varepsilon y) \frac{(\boldsymbol{b}_t(x - \varepsilon y) - \boldsymbol{b}_t(x)) \cdot \nabla\rho(y)}{\varepsilon} \, dy - (w\,\mathrm{div}\,\boldsymbol{b}_t) * \rho_\varepsilon(x).$$

$$(4.6)$$

Introducing the commutators in the (easier) conservative form

$$R^\varepsilon := (D_x \cdot (\boldsymbol{b}w)) * \rho_\varepsilon - D_x \cdot (\boldsymbol{b}w^\varepsilon)$$

(here we set again $w^\varepsilon := w * \rho_\varepsilon$) it suffices to show that $R^\varepsilon = L^\varepsilon - w^\varepsilon\,\mathrm{div}\,\boldsymbol{b}_t$, where

$$L^\varepsilon(t,x) := \int_{\mathbb{R}^d} w(t,z)(\boldsymbol{b}_t(x) - \boldsymbol{b}_t(z)) \cdot \nabla\rho_\varepsilon(z - x) \, dz.$$

Indeed, for any test function φ, we have that $\langle R^\varepsilon, \varphi \rangle$ is given by

$$-\int_I \int w\boldsymbol{b} \cdot \nabla\rho_\varepsilon * \varphi\, dy dt - \int_I \int \varphi\boldsymbol{b} \cdot \nabla\rho_\varepsilon * w\, dx dt - \int_I \int w^\varepsilon \varphi\,\mathrm{div}\,\boldsymbol{b}_t dt$$

$$= -\int_I \int \int w_t(y)\boldsymbol{b}_t(y) \cdot \nabla\rho_\varepsilon(y - x)\varphi(x)\, dx dy dt$$

$$-\int_I \int \int \boldsymbol{b}_t(x)\nabla\rho_\varepsilon(x - y)w_t(y)\varphi(x)\, dy dx dt - \int_I \int w^\varepsilon \varphi\,\mathrm{div}\,\boldsymbol{b}_t\, dx dt$$

$$= \int_I \int L^\varepsilon \varphi\, dx dt - \int_I \int w^\varepsilon\,\mathrm{div}\,\boldsymbol{b}_t\, dx dt$$

(in the last equality we used the fact that $\nabla\rho$ is odd).

Then, one uses the strong convergence of translations in L^p and the strong convergence of the difference quotients (a property that *characterizes* functions in Sobolev spaces)

$$\frac{u(x + \varepsilon z) - u(x)}{\varepsilon} \to \nabla u(x)z \qquad \text{strongly in } L^1_{\mathrm{loc}}, \text{ for } u \in W^{1,1}_{\mathrm{loc}}$$

to obtain that r^ε strongly converge in $L^1_{\mathrm{loc}}(I \times \mathbb{R}^d)$ to

$$-w(t,x)\int_{\mathbb{R}^d} \langle \nabla\boldsymbol{b}_t(x)y, \nabla\rho(y) \rangle \, dy - w(t,x)\mathrm{div}\,\boldsymbol{b}_t(x).$$

The elementary identity

$$\int_{\mathbb{R}^d} y_i \frac{\partial\rho}{\partial y_j} \, dy = -\delta_{ij}$$

then shows that the limit is 0 (this can also be derived by the fact that, in any case, the limit of r^ε in the distribution sense should be 0).

In this context, given $\bar{\mu} = \rho \mathcal{L}^d$ with $\rho \in L^1 \cap L^\infty$, the $\mathcal{L}$-Lagrangian flow starting from $\bar{\mu}$ (at time 0) is defined by the following two properties:

(a) $\boldsymbol{X}(\cdot, x)$ is absolutely continuous in $[0, T]$ and satisfies

$$\boldsymbol{X}(t, x) = x + \int_0^t \boldsymbol{b}_s(\boldsymbol{X}(s, x)) \, ds \qquad \forall t \in [0, T]$$

for $\bar{\mu}$-a.e. x;

(b) $\boldsymbol{X}(t, \cdot)_{\#} \bar{\mu} \leq C \mathcal{L}^d$ for all $t \in [0, T]$, with C independent of t.

Summing up what we obtained so far, the general theory provides us with the following existence and uniqueness result.

Theorem 4.3 (Existence and Uniqueness of $\mathcal{L}$-Lagrangian Flows). *Let $\boldsymbol{b} \in L^1\left([0, T]; W^{1,1}_{\mathrm{loc}}(\mathbb{R}^d; \mathbb{R}^d)\right)$ be satisfying*

(i) $\dfrac{|\boldsymbol{b}|}{1 + |x|} \in L^1\left([0, T]; L^1(\mathbb{R}^d)\right) + L^1\left([0, T]; L^\infty(\mathbb{R}^d)\right);$

(ii) $[\mathrm{div}\, \boldsymbol{b}_t]^- \in L^1\left([0, T]; L^\infty(\mathbb{R}^d)\right).$

Then the $\mathcal{L}$-Lagrangian flow relative to $\boldsymbol{b}$ exists and is unique.

Proof. By the previous results, the comparison principle holds for the continuity equation relative to $\boldsymbol{b}$. Therefore the general theory previously developed applies, and Theorem 3.3 provides *uniqueness* of the $\mathcal{L}$-Lagrangian flow.

As for the *existence*, still the general theory (Theorem 3.5) tells us that it can be achieved provided we are able to solve, within $\mathcal{L}$, the continuity equation

$$\frac{d}{dt} w + D_x \cdot (\boldsymbol{b} w) = 0 \tag{4.7}$$

for any nonnegative initial datum $w_0 \in L^1 \cap L^\infty$. The existence of these solutions can be immediately achieved by a smoothing argument: we approximate $\boldsymbol{b}$ in L^1_{loc} by smooth $\boldsymbol{b}^h$ with a uniform bound in $L^1(L^\infty)$ for $[\mathrm{div}\, \boldsymbol{b}^h_t]^-$. This bound, in turn, provides a uniform lower bound on $J\boldsymbol{X}^h$ and finally a uniform upper bound on $w^h_t = (w_0 / J\boldsymbol{X}^h_t) \circ (\boldsymbol{X}^h_t)^{-1}$, solving

$$\frac{d}{dt} w^h + D_x \cdot (\boldsymbol{b}^h w^h) = 0.$$

Therefore, any weak limit of w^h solves (4.7).

Notice also that, choosing for instance a Gaussian, we obtain that the $\mathcal{L}$-Lagrangian flow is well defined up to $\mathcal{L}^d$-negligible sets (and independent of $\bar{\mu} \ll \mathcal{L}^d$, thanks to Remark 3.2).

It is interesting to compare our characterization of Lagrangian flows with the one given in [53]. Heuristically, while the Di Perna-Lions one is based on the semigroup of transformations $x \mapsto \boldsymbol{X}(t, x)$, our one is based on the properties of the map $x \mapsto \boldsymbol{X}(\cdot, x)$.

Remark 4.3. The definition of the flow in [53] is based on the following three properties:

(a) $\dfrac{\partial \boldsymbol{Y}}{\partial t}(t, s, x) = b\left(t, \boldsymbol{Y}(t, s, x)\right)$ and $\boldsymbol{Y}(s, s, x) = x$ in the distribution sense in $(0, T) \times \mathbb{R}^d$;

(b) the image λ_t of $\mathcal{L}^d$ under $\boldsymbol{Y}(t, s, \cdot)$ satisfies

$$\frac{1}{C} \mathcal{L}^d \leq \lambda_t \leq C \mathcal{L}^d \qquad \text{for some constant } C > 0;$$

(c) for all s, $s', t \in [0, T]$ we have

$$\boldsymbol{Y}\left(t, s, \boldsymbol{Y}(s, s', x)\right) = \boldsymbol{Y}(t, s', x) \qquad \text{for } \mathcal{L}^d\text{-a.e. } x.$$

Then, $\boldsymbol{Y}(t, s, x)$ corresponds, in our notation, to the flow $\boldsymbol{X}^s(t, x)$ starting at time s (well defined even for $t < s$ if one has two-sided L^∞ bounds on the divergence).

In our setting condition (c) can be recovered as a consequence with the following argument: assume to fix the ideas that $s' \leq s \leq T$ and define

$$\tilde{\boldsymbol{X}}(t, x) := \begin{cases} \boldsymbol{X}^{s'}(t, x) & \text{if } t \in [s', s]; \\ \boldsymbol{X}^s\left(t, \boldsymbol{X}^{s'}(s, x)\right) & \text{if } t \in [s, T]. \end{cases}$$

It is immediate to check that $\tilde{\boldsymbol{X}}(\cdot, x)$ is an integral solution of the ODE in $[s', T]$ for $\mathcal{L}^d$-a.e. x and that $\tilde{\boldsymbol{X}}(t, \cdot)_\# \bar{\mu}$ is bounded by $C^2 \mathcal{L}^d$. Then, Theorem 4.3 (with s' as initial time) gives $\tilde{\boldsymbol{X}}(\cdot, x) = \boldsymbol{X}(\cdot, s', x)$ in $[s', T]$ for $\mathcal{L}^d$-a.e. x, whence (c) follows.

Moreover, the stability Theorem 3.6 can be read in this context as follows. We state it for simplicity only in the case of equi-bounded vectorfields (see [9] for more general results).

Theorem 4.4 (Stability). *Let $\boldsymbol{b}^h$, $\boldsymbol{b} \in L^1\left([0, T]; W^{1,1}_{\text{loc}}(\mathbb{R}^d; \mathbb{R}^d)\right)$, let $\boldsymbol{X}^h$, $\boldsymbol{X}$ be the $\mathcal{L}$-Lagrangian flows relative to $\boldsymbol{b}^h$, $\boldsymbol{b}$, let $\bar{\mu} = \rho \mathcal{L}^d \in \mathcal{M}_+(\mathbb{R}^d)$ and assume that*

(i) $\boldsymbol{b}^h \to \boldsymbol{b}$ in $L^1_{\text{loc}}\left((0, T) \times \mathbb{R}^d\right)$;
(ii) $|\boldsymbol{b}_h| \leq C$ for some constant C independent of h;
(iii) $[\text{div } \boldsymbol{b}^h_t]^-$ is bounded in $L^1\left([0, T]; L^\infty(\mathbb{R}^d)\right)$.

Then,

$$\lim_{h \to \infty} \int_{\mathbb{R}^d} \max_{[0, T]} |\boldsymbol{X}^h(\cdot, x) - \boldsymbol{X}(\cdot, x)| \wedge \rho(x) \, dx = 0.$$

Proof. It is not restrictive, by an approximation argument, to assume that ρ has a compact support. Under this assumption, (i) and (iii) ensure that $\mu^h_t \leq M \chi_{B_R} \mathcal{L}^d$ for some constants M and R independent of h and t. Denoting by μ_t the weak limit of μ_t^h, choosing $\Theta(z) = |z|^2$ in (iii) of Theorem 3.6, we have to check that

$$\lim_{h \to \infty} \int_0^T \int_{\mathbb{R}^d} \frac{|\boldsymbol{b}_h|^2}{1 + |x|} \, d\mu_t^h dt = \int_0^T \int_{\mathbb{R}^d} \frac{|\boldsymbol{b}|^2}{1 + |x|} \, d\mu_t dt. \qquad (4.8)$$

Let $\varepsilon > 0$ and let $B \subset (0, T) \times B_R$ be an open set given by Egorov theorem, such that $\boldsymbol{b}_h \to \boldsymbol{b}$ uniformly on $[0, T] \times B_R \setminus B$ and $\mathcal{L}^{d+1}(B) < \varepsilon$. Let also $\tilde{\boldsymbol{b}}_\varepsilon$ be such that $|\tilde{\boldsymbol{b}}_\varepsilon| \leq C$ and $\tilde{\boldsymbol{b}}_\varepsilon = \boldsymbol{b}$ on $[0, T] \times B_R \setminus B$. We write

$$\int_0^T \int_{\mathbb{R}^d} \frac{|\boldsymbol{b}_h|^2}{1+|x|} \, d\mu_t^h dt - \int_0^T \int_{\mathbb{R}^d} \frac{|\tilde{\boldsymbol{b}}_\varepsilon|^2}{1+|x|} \, d\mu_t^h dt =$$

$$= \int_{[0,T] \times B_R \setminus B} \frac{|\boldsymbol{b}_h|^2 - |\tilde{\boldsymbol{b}}_\varepsilon|^2}{1+|x|} \, d\mu_t^h dt + \int_B \frac{|\boldsymbol{b}_h|^2 - |\tilde{\boldsymbol{b}}_\varepsilon|^2}{1+|x|} \, d\mu_t^h dt,$$

so that

$$\limsup_{h \to \infty} \left| \int_0^T \int_{\mathbb{R}^d} \frac{|\boldsymbol{b}_h|^2}{1 + |x|} \, d\mu_t^h dt - \int_0^T \int_{\mathbb{R}^d} \frac{|\tilde{\boldsymbol{b}}_\varepsilon|^2}{1 + |x|} \, d\mu_t dt \right| \leq 2C^2 M \varepsilon.$$

As ε is arbitrary and

$$\lim_{\varepsilon \to 0} \int_0^T \int_{\mathbb{R}^d} \frac{|\tilde{\boldsymbol{b}}_\varepsilon|^2}{1 + |x|} \, d\mu_t dt = \int_0^T \int_{\mathbb{R}^d} \frac{|\boldsymbol{b}|^2}{1 + |x|} \, d\mu_t dt$$

this proves that (4.8) is fulfilled.

Finally, we conclude this section with the illustration of some recent results [64], [13], [14] that seem to be more specific of the Sobolev case, concerned with the "differentiability" w.r.t. to x of the flow $\boldsymbol{X}(t, x)$. These results provide a sort of bridge with the standard Cauchy-Lipschitz calculus:

Theorem 4.5. *There exist Borel maps $L_t : \mathbb{R}^d \to M^{d \times d}$ satisfying*

$$\lim_{h \to 0} \frac{\boldsymbol{X}(t, x + h) - \boldsymbol{X}(t, x) - L_t(x)h}{|h|} = 0 \quad \text{locally in measure}$$

for any $t \in [0, T]$. If, in addition, we assume that

$$\int_0^T \int_{B_R} |\nabla \boldsymbol{b}_t| \ln(2 + |\nabla \boldsymbol{b}_t|) \, dx dt < +\infty \qquad \forall R > 0$$

then the flow has the following "local" Lipschitz property: for any $\varepsilon > 0$ there exists a Borel set A with $\bar{\mu}(\mathbb{R}^d \setminus A) < \varepsilon$ such that $\boldsymbol{X}(t, \cdot)|_A$ is Lipschitz for any $t \in [0, T]$.

According to this result, L can be thought as a (very) weak derivative of the flow $\boldsymbol{X}$. It is still not clear whether the local Lipschitz property holds in the $W_{\text{loc}}^{1,1}$ case, or in the BV_{loc} case discussed in the next section.

5 Vector Fields with a BV Spatial Regularity

In this section we prove the renormalization Theorem 4.2 under the weaker assumption of a BV dependence w.r.t. the spatial variables, but still assuming that

$$D \cdot b_t \ll \mathcal{L}^d \qquad \text{for } \mathcal{L}^1\text{-a.e. } t \in (0, T). \tag{5.1}$$

Theorem 5.1. Let $b \in L^1_{\text{loc}}\left((0, T); BV_{\text{loc}}(\mathbb{R}^d; \mathbb{R}^d)\right)$ be satisfying (5.1). Then any distributional solution $w \in L^\infty_{\text{loc}}\left((0, T) \times \mathbb{R}^d\right)$ of

$$\frac{d}{dt} w + D_x \cdot (bw) = c \in L^1_{\text{loc}}\left((0, T) \times \mathbb{R}^d\right)$$

is a renormalized solution.

We try to give reasonably detailed proof of this result, referring to the original paper [7] for minor details. Before doing that we set up some notation, denoting by $Db_t = D^a b_t + D^s b_t = \nabla b_t \mathcal{L}^d + D^s b_t$ the Radon–Nikodym decomposition of Db_t in absolutely continuous and singular part w.r.t. $\mathcal{L}^d$. We also introduce the measures $|Db|$ and $|D^s b|$ by integration w.r.t. the time variable, i.e.

$$\int \varphi(t, x) \, d|Db| := \int_0^T \int_{\mathbb{R}^d} \varphi(t, x) \, d|Db_t| \, dt,$$

$$\int \varphi(t, x) \, d|D^s b| := \int_0^T \int_{\mathbb{R}^d} \varphi(t, x) \, d|D^s b_t| \, dt.$$

We shall also assume, by the locality of the arguments involved, that $\|w\|_\infty \leq 1$.

We are going to find two estimates on the commutators, quite sensitive to the choice of the convolution kernel, and then combine them in a (pointwise) kernel optimization argument.

Step 1 (Anisotropic Estimate). Let us start from the expression

$$r^\varepsilon(t, x) = \int_{\mathbb{R}^d} w(t, x - \varepsilon y) \frac{(b_t(x - \varepsilon y) - b_t(x)) \cdot \nabla \rho(y)}{\varepsilon} \, dy - (w \operatorname{div} b_t) * \rho_\varepsilon(x) \tag{5.2}$$

of the commutators $(b \cdot \nabla w) * \rho_\varepsilon - b \cdot (\nabla (w * \rho_\varepsilon))$: since $b_t \notin W^{1,1}$ we cannot use anymore the strong convergence of the difference quotients. However, for any function $u \in BV_{\text{loc}}$ and any $z \in \mathbb{R}^d$ with $|z| < \varepsilon$ we have a classical L^1 estimate on the difference quotients

$$\int_K |u(x + z) - u(x)| \, dx \leq |D_z u|(K_\epsilon) \quad \text{for any } K \subset \mathbb{R}^d \text{ compact,}$$

where $Du = (D_1 u, \dots, D_d u)$ stands for the distributional derivative of u, $D_z u = \langle Du, z \rangle = \sum_i z_i D_i u$ denotes the component along z of Du and K_ε is the open ε-neighbourhood of K. Its proof follows from an elementary smoothing and lower semicontinuity argument.

We notice that, setting $Db_t = M_t|Db_t|$, we have

$$D_z\langle b_t, \nabla\rho(z)\rangle = \langle M_t(\cdot)z, \nabla\rho(z)\rangle|Db| \qquad \forall z \in \mathbb{R}^d$$

and therefore the L^1 estimate on difference quotients gives the *anisotropic estimate*

$$\limsup_{\varepsilon\downarrow 0} \int_K |r^\varepsilon|\,dx \le \int_K \int_{\mathbb{R}^d} |\langle M_t(x)z, \nabla\rho(z)\rangle|\,dz\,d|Db|(t,x) + d|D^a b|(K) \tag{5.3}$$

for any compact set $K \subset (0,T) \times \mathbb{R}^d$.

Step 2 (Isotropic Estimate). On the other hand, a different estimate of the commutators that reduces to the standard one when $b(t,\cdot) \in W^{1,1}_{\mathrm{loc}}$ can be achieved as follows. Let us start from the case $d = 1$: if μ is a $\mathbb{R}^m$-valued measure in $\mathbb{R}$ with locally finite variation, then by Jensen's inequality the functions

$$\hat{\mu}_\varepsilon(t) := \frac{\mu([t, t+\varepsilon])}{\varepsilon} = \mu * \frac{\chi_{[-\varepsilon,0]}}{\varepsilon}(t), \qquad t \in \mathbb{R}$$

satisfy

$$\int_K |\hat{\mu}_\varepsilon|\,dt \le |\mu|(K_\varepsilon) \quad \text{for any compact set } K \subset \mathbb{R}, \tag{5.4}$$

where K_ε is again the open ε neighbourhood of K. A density argument based on (5.4) then shows that $\hat{\mu}_\varepsilon$ converge in $L^1_{\mathrm{loc}}(\mathbb{R})$ to the density of μ with respect to $\mathcal{L}^1$ whenever $\mu \ll \mathcal{L}^1$. If $u \in BV_{\mathrm{loc}}$ and $\varepsilon > 0$ we know that

$$\frac{u(x+\varepsilon) - u(x)}{\varepsilon} = \frac{Du([x, x+\varepsilon])}{\varepsilon} = \frac{D^a u([x, x+\varepsilon])}{\varepsilon} + \frac{D^s u([x, x+\varepsilon])}{\varepsilon}$$

for $\mathcal{L}^1$-a.e. x (the exceptional set possibly depends on ε). In this way we have canonically split the difference quotient of u as the sum of two functions, one *strongly* converging to ∇u in L^1_{loc}, and the other one having an L^1 norm on any compact set K asymptotically smaller than $|D^s u|(K)$.

If we fix the direction z of the difference quotient, the slicing theory of BV functions gives that this decomposition can be carried on also in d dimensions, showing that the difference quotients

$$\frac{b_t(x+\varepsilon z) - b_t(x)}{\varepsilon}$$

can be canonically split into two parts, the first one strongly converging in $L^1_{\mathrm{loc}}(\mathbb{R}^d)$ to $\nabla b_t(x)z$, and the second one having an L^1 norm on K asymptotically smaller than $|\langle D^s b_t, z\rangle|(K)$. Then, repeating the DiPerna–Lions argument and taking into account the error induced by the presence of the second part of the difference quotients, we get the *isotropic estimate*

$$\limsup_{\varepsilon \downarrow 0} \int_K |r^\varepsilon|\, dx \le \left(\int_K \int_{\mathbb{R}^d} |z||\nabla \rho(z)|\, dz \right) d|D^s b|(t,x) \qquad (5.5)$$

for any compact set $K \subset (0,T) \times \mathbb{R}^d$.

Step 3 (Reduction to a Pointwise Optimization Problem). Roughly speaking, the isotropic estimate is useful in the regions where the absolutely continuous part is the dominant one, so that $|D^s b|(K) << |D^a b|(K)$, while the anisotropic one turns out to be useful in the regions where the dominant part is the singular one, i.e. $|D^a b|(K) << |D^s b|(K)$. Since the two measures are mutually singular, for a typical small ball K only one of these two situations occurs. Let us see how the two estimates can be combined: coming back to the smoothing scheme, we have

$$\frac{d}{dt}\beta(w^\varepsilon) + \boldsymbol{b} \cdot \nabla\beta(w^\varepsilon) - \beta'(w^\varepsilon)c * \rho_\varepsilon = \beta'(w^\varepsilon)r^\varepsilon. \qquad (5.6)$$

Let L be the supremum of $|\beta'|$ on $[-1,1]$. Then, since K is an arbitrary compact set, (5.5) tells us that any limit measure ν of $|\beta'(w^\varepsilon)r^\varepsilon|\mathcal{L}^d$ as $\varepsilon \downarrow 0$ satisfies

$$\nu \le LI(\rho)|D^s b| \quad \text{with} \quad I(\rho) := \int_{\mathbb{R}^d} |z||\nabla \rho(z)|\, dz.$$

and, in particular, is singular with respect to $\mathcal{L}^d$. On the other hand, the estimate (5.3) tells also us that

$$\nu \le L \int_{\mathbb{R}^d} |\langle M.(\cdot)z, \nabla \rho(z)\rangle|\, dz|Db| + d|D^a b|(K).$$

The second estimate and the singularity of ν with respect to $\mathcal{L}^d$ give

$$\nu \le L \int_{\mathbb{R}^d} |\langle M.(\cdot)z, \nabla \rho(z)\rangle|\, dz|D^s b|. \qquad (5.7)$$

Notice that in this way we got rid of the potentially dangerous term $I(\rho)$: in fact, we are going to choose *very* anisotropic kernels ρ on which $I(\rho)$ can be arbitrarily large. The measure ν can of course depend on the choice of ρ, but (5.6) tells us that the "defect" measure

$$\sigma := \frac{d}{dt}\beta(w_t) + \boldsymbol{b} \cdot \nabla\beta(w_t) - c_t \beta'(w_t),$$

clearly independent of ρ, satisfies $|\sigma| \le \nu$. Eventually we obtain

$$|\sigma| \le L\Lambda(M.(\cdot), \rho)|D^s b| \quad \text{with} \quad \Lambda(N, \rho) := \int_{\mathbb{R}^d} |\langle Nz, \nabla \rho(z)\rangle|\, dz. \qquad (5.8)$$

For (x,t) *fixed*, we are thus led to the minimum problem

$$G(N) := \inf\left\{ \Lambda(N,\rho) : \rho \in C_c^\infty(B_1),\ \rho \ge 0,\ \int_{\mathbb{R}^d} \rho = 1 \right\} \qquad (5.9)$$

with $N = M_t(x)$. Indeed, notice that (5.8) gives

$$|\sigma| \leq L \inf_{\rho \in D} \Lambda(M.(\cdot), \rho)|D^s b|$$

for any countable set D of kernels ρ, and the continuity of $\rho \mapsto \Lambda(N, \rho)$ w.r.t. the $W^{1,1}(B_1)$ norm and the separability of $W^{1,1}(B_1)$ give

$$|\sigma| \leq L G(M.(\cdot))|D^s b|. \tag{5.10}$$

Notice now that the assumption that $D \cdot b_t \ll \mathcal{L}^d$ for $\mathcal{L}^1$-a.e. $t \in (0, T)$ gives

$$\text{trace } M_t(x)|D^s b_t| = 0 \qquad \text{for } \mathcal{L}^1\text{-a.e. } t \in (0, T).$$

Hence, recalling the definition of $|D^s b|$, the trace of $M_t(x)$ vanishes for $|D^s b|$-a.e. (t, x). Applying the following lemma, a courtesy of Alberti, and using (5.10) we obtain that $\sigma = 0$, thus concluding the proof.

Lemma 5.1 (Alberti). *For any $d \times d$ matrix N the infimum in (5.9) is $|\text{trace } N|$.*

Proof. We have to build kernels ρ in such a way that the field Nz is as much tangential as possible to the level sets of ρ. Notice first that the lower bound follows immediately by the identity

$$\int_{\mathbb{R}^d} \langle Nz, \nabla\rho(z)\rangle \, dz = \int_{\mathbb{R}^d} -\rho(z)\text{div } Nz + \text{div }(\rho(z)Nz) \, dz = -\text{trace } N.$$

Hence, we have to show only the upper bound. Again, by the identity

$$\langle Nz, \nabla\rho(z)\rangle = \text{div }(Nz\rho(z)) - \text{trace } N\rho(z)$$

it suffices to show that for any $T > 0$ there exists ρ such that

$$\int_{\mathbb{R}^d} |\text{div }(Nz\rho(z))| \, dz \leq \frac{2}{T}. \tag{5.11}$$

The heuristic idea is (again...) to build ρ as the superposition of elementary probability measures associated to the curves $e^{tN}x$, $0 \leq t \leq T$, on which the divergence operator can be easily estimated. Given a smooth convolution kernel θ with compact support, it turns out that the function

$$\rho(z) := \frac{1}{T}\int_0^T \theta(e^{-tN}z)e^{-t\,\text{trace } N} \, dt \tag{5.12}$$

has the required properties (here $e^{tN}x = \sum_i t^i N^i x / i!$ is the solution of the ODE $\dot{\gamma} = N\gamma$ with the initial condition $\gamma(0) = x$). Indeed, it is immediate to check that ρ is smooth and compactly supported. To estimate the divergence of $Nz\rho(z)$, we notice that $\rho = \int \theta(x)\mu_x \, dx$, where μ_x are the probability

1-dimensional measures concentrated on the image of the curves $t \mapsto e^{tN}x$ defined by

$$\mu_x := (e^{\cdot N}x)_\# (\frac{1}{T}\mathcal{L}^1 \llcorner [0,T]).$$

Indeed, for any $\varphi \in C_c^\infty(\mathbb{R}^d)$ we have

$$\int_{\mathbb{R}^d} \theta(x)\langle \mu_x, \varphi \rangle \, dx = \frac{1}{T} \int_0^T \int_{\mathbb{R}^d} \theta(x)\varphi(e^{tN}x) \, dx dt$$

$$= \frac{1}{T} \int_0^T \int_{\mathbb{R}^d} \theta(e^{-tN}y)e^{-t\,\mathrm{trace}N} \varphi(y) \, dy dt$$

$$= \int_{\mathbb{R}^d} \rho(y)\varphi(y) \, dy.$$

By the linearity of the divergence operator, it suffices to check that

$$|D_z \cdot (Nz\mu_x)|(\mathbb{R}^d) \le \frac{2}{T} \qquad \forall x \in \mathbb{R}^d.$$

But this is elementary, since

$$\int_{\mathbb{R}^d} \langle Nz, \nabla\varphi(z) \rangle \, d\mu_x(z) = \frac{1}{T} \int_0^T \langle Ne^{tN}x, \nabla\varphi(e^{tN}x) \rangle \, dt = \frac{\varphi(e^{TN}x) - \varphi(x)}{T}$$

for any $\varphi \in C_c^\infty(\mathbb{R}^d)$, so that $TD_z \cdot (Nz\mu_x) = \delta_x - \delta_{e^{TN}x}$.

The original argument in [7] was slightly different and used, instead of Lemma 5.1, a much deeper result, still due to Alberti, saying that for a BV_{loc} function $u : \mathbb{R}^d \to \mathbb{R}^m$ the matrix $M(x)$ in the polar decomposition $Du = M|Du|$ has rank 1 for $|D^s u|$-a.e. x, i.e. there exist unit vectors $\xi(x) \in \mathbb{R}^d$ and $\eta(x) \in \mathbb{R}^m$ such that $M(x)z = \eta(x)\langle z, \xi(x) \rangle$. In this case the asymptotically optimal kernels are much easier to build, by mollifying in the ξ direction much faster than in all other ones. This is precisely what Bouchut and Lions did in some particular cases (respectively "Hamiltonian" vector fields and piecewise Sobolev ones).

As in the Sobolev case we can now obtain from the general theory given in Section 3 *existence and uniqueness* of $\mathcal{L}$-Lagrangian flows, with $\mathcal{L} = L^\infty(L^1) \cap L^\infty(L^\infty)$: we just replace in the statement of Theorem 4.3 the assumption $b \in L^1\left([0,T]; W^{1,1}_{\mathrm{loc}}(\mathbb{R}^d; \mathbb{R}^d)\right)$ with $b \in L^1\left([0,T]; BV_{\mathrm{loc}}(\mathbb{R}^d; \mathbb{R}^d)\right)$, assuming as usual that $D \cdot b_t \ll \mathcal{L}^d$ for $\mathcal{L}^1$-a.e. $t \in [0,T]$.

Analogously, with the same replacements in Theorem 4.4 (for b and b^h) we obtain *stability* of $\mathcal{L}$-Lagrangian flows.

6 Applications

6.1. A System of Conservation Laws. Let us consider the Cauchy problem (studied in one space dimension by Keyfitz–Kranzer in [63])

$$\frac{d}{dt}u + \sum_{i=1}^{d} \frac{\partial}{\partial x_i} \left(\boldsymbol{f}_i(|u|)u \right) = 0, \qquad u : \mathbb{R}^d \times (0, +\infty) \to \mathbb{R}^k \qquad (6.1)$$

with the initial condition $u(\cdot, 0) = \bar{u}$. Here $\boldsymbol{f} : \mathbb{R} \to \mathbb{R}^d$ is a C^1 function.

In a recent paper [32] Bressan showed that the problem can be ill-posed for L^∞ initial data and he conjectured that it could be well posed for BV initial data, suggesting to extend to this case the classical method of characteristics. In [8] we proved that this procedure can really be implemented, thanks to the results in [7], for initial data $\bar{u}$ such that $\bar{\rho} := |\bar{u}| \in BV \cap L^\infty$, with $1/|\bar{u}| \in L^\infty$. Later on, in a joint work with Bouchut and De Lellis [10], we proved that the lower bound on $\bar{\rho}$ is not necessary and, moreover, we proved that the solution built in [8] is unique in a suitable class of admissible functions: those whose modulus ρ satisfies the scalar PDE

$$\frac{d}{dt}\rho + \sum_{i=1}^{d} \frac{\partial}{\partial x_i} \left(\boldsymbol{f}_i(\rho)\rho \right) = 0 \qquad (6.2)$$

in the Kruzhkov sense (i.e. $\eta(\rho)_t + D_x \cdot (\boldsymbol{q}(\rho)) \leq 0$ for any convex entropy-entropy flux pair $(\eta, \boldsymbol{q})$, here $(s\boldsymbol{f})'(s)\eta'(s) = \boldsymbol{q}'(s))$, with the initial condition $\rho(0, \cdot) = \bar{\rho}$.

Notice that the regularity theory for this class of solutions gives that $\rho \in L^\infty \cap BV_{\mathrm{loc}} \left([0, +\infty) \times \mathbb{R}^d \right)$, due to the BV regularity and the boundedness of $|\bar{u}|$. Furthermore the maximum principle gives $0 < 1/\rho \leq 1/|\bar{u}| \in L^\infty$.

In order to obtain the (or, better, a) solution u we can *formally* decouple the system, writing

$$u = \theta\rho, \qquad \bar{u} = \bar{\theta}\bar{\rho}, \qquad |\theta| = |\bar{\theta}| = 1,$$

thus reducing the problem to the system (decoupled, if one neglects the constraint $|\theta| = 1$) of transport equations

$$\theta_t + \sum_{i=1}^{d} \frac{\partial}{\partial x_i} \left(\boldsymbol{f}_i(\rho)\theta \right) = 0 \qquad (6.3)$$

with the initial condition $\theta(0, \cdot) = \bar{\theta}$.

A *formal* solution of the system, satisfying also the constraint $|\theta| = 1$, is given by

$$\theta(t, x) := \bar{\theta} \left([\boldsymbol{X}(t, \cdot)]^{-1}(x) \right),$$

where $\boldsymbol{X}(t, \cdot)$ is the flow associated to $\boldsymbol{f}(\rho)$. Notice that the non-autonomous vector field $\boldsymbol{f}(\rho)$ is bounded and of class BV_{loc}, but the theory illustrated in these lectures is not immediately applicable because its divergence *is not* absolutely continuous with respect to $\mathcal{L}^{d+1}$. In this case, however, a simple argument still allows the use of the theory, representing $\boldsymbol{f}(\rho)$ as a part of the autonomous vector field $\boldsymbol{b} := (\rho, \rho\boldsymbol{f}(\rho))$ in $\mathbb{R}^+ \times \mathbb{R}^d$. This new vector field is still BV_{loc} and bounded, and it is divergence-free due to (6.2).

At this point, it is not hard to see that the reparameterization of the flow $(t(s), \boldsymbol{x}(s))$ associated to $\boldsymbol{b}$

$$(\dot{t}(s), \dot{\boldsymbol{x}}(s)) = (\rho(t(s), \boldsymbol{x}(s)), \boldsymbol{f}(\rho(t(s), \boldsymbol{x}(s)))\rho(t(s), \boldsymbol{x}(s)))$$

defined by $\tilde{\boldsymbol{x}}(t) = \boldsymbol{x}(t(s)^{-1}(t))$ (and here we use the assumption $\rho > 0$) defines a flow for the vector field $\boldsymbol{f}(\rho)$ we were originally interested to.

In this way we get a kind of formal, or pointwise, solution of the system (6.2), that could indeed be very far from being a *distributional* solution.

But here comes into play the stability theorem, showing that all formal computations above can be justified just assuming first $(\rho, \boldsymbol{f}(\rho))$ smooth, and then by approximation (see [8] for details).

6.2. Lagrangian Solutions of Semi-Geostrophic Equations. The semi-geostrophic equations are a simplifies model of the atmosphere/ocean flows [45], described by the system of transport equations

(SGE)
$$\begin{cases} \dfrac{d}{dt}\partial_2 p + \boldsymbol{u} \cdot \nabla \partial_2 p = -u_2 + \partial_1 p \\[2mm] \dfrac{d}{dt}\partial_1 p + \boldsymbol{u} \cdot \nabla \partial_1 p = -u_1 - \partial_2 p \\[2mm] \dfrac{d}{dt}\partial_3 p + \boldsymbol{u} \cdot \nabla \partial_3 p = 0. \end{cases}$$

Here $\boldsymbol{u}$, the velocity, is a divergence-free field, p is the pressure and $\rho := -\partial_3 p$ represents the density of the fluid. We consider the problem in $[0, T] \times \Omega$, with Ω bounded and convex. Initial conditions are given on the pressure and a no-flux condition through $\partial\Omega$ is imposed for all times.

Introducing the modified pressure $P_t(x) := p_t(x) + (x_1^2 + x_2^2)/2$, (SGE) can be written in a more compact form as

$$\frac{d}{dt}\nabla P + \boldsymbol{u} \cdot \nabla^2 P = J(\nabla P - x) \quad \text{with} \quad J := \begin{pmatrix} 0 & -1 & 0 \\ 1 & 0 & 0 \\ 0 & 0 & 0 \end{pmatrix}. \tag{6.4}$$

Existence (and uniqueness) of solutions are still open for this problem. In [20] and [46], existence results have been obtained in the so-called dual coordinates, where we replace the physical variable x by $X = \nabla P_t(x)$. Under this change of variables, *and assuming P_t to be convex*, the system becomes

$$\frac{d}{dt}\alpha_t + D_x \cdot (U_t\alpha_t) = 0 \quad \text{with} \quad U_t(X) := J(X - \nabla P_t^*(X)) \tag{6.5}$$

with $\alpha_t := (\nabla P_t)_\#(\mathcal{L}_\Omega)$ (here we denote by $\mathcal{L}_\Omega$ the restriction of $\mathcal{L}^d$ to Ω). Indeed, for any test function φ we can use the fact that $\boldsymbol{u}$ is divergence-free to obtain:

$$\frac{d}{dt}\int_{\mathbb{R}^d}\varphi\,d\alpha_t = \int_{\mathbb{R}^d}\nabla\varphi(\nabla P_t)\cdot\frac{d}{dt}\nabla P_t\,dx$$

$$= \int_{\mathbb{R}^d}\nabla\varphi(\nabla P_t)\cdot J(\nabla P_t - x)\,dx + \int_{\mathbb{R}^d}\nabla\varphi(\nabla P_t)\nabla^2 P_t\cdot\boldsymbol{u}\,dx$$

$$= \int_{\mathbb{R}^d}\nabla\varphi\cdot J(X - \nabla P_t^*)\,d\alpha_t + \int_{\mathbb{R}^d}\nabla(\varphi\circ\nabla P_t)\cdot\boldsymbol{u}\,dx$$

$$= \int_{\mathbb{R}^d}\nabla\varphi\cdot\boldsymbol{U}_t\,d\alpha_t.$$

Existence of a solution to (6.5) can be obtained by a suitable time discretization scheme. Now the question is: can we go back to the original physical variables ? An important step forward has been achieved by Cullen and Feldman in [47], with the concept of *Lagrangian* solution of (SGE).

Taking into account that the vector field $\boldsymbol{U}_t(X) = J(X - \nabla P_t^*(X))$ is BV, bounded and divergence-free, there is a well defined, stable and measure preserving flow $\boldsymbol{X}(t, X) = \boldsymbol{X}_t(X)$ relative to $\boldsymbol{U}$. This flow can be carried back to the physical space with the transformation

$$F_t(x) := \nabla P_t^* \circ \boldsymbol{X}_t \circ \nabla P_0(x),$$

thus defining maps F_t preserving $\mathcal{L}_\Omega^d$.

Using the stability theorem can also show that $Z_t(x) := \nabla P_t(F_t(x))$ solve, in the distributions sense, the Lagrangian form of (6.4), i.e.

$$\frac{d}{dt}Z_t(x) = J(Z_t - F_t) \qquad (6.6)$$

This provides us with a sort of weak solution of (6.4), and it is still an open problem how the Eulerian form could be recovered (see Section 7).

7 Open Problems, Bibliographical Notes, and References

Section 2. The material contained in this section is classical. Good references are [56], Chapter 8 of [12], [29] and [53]. For the proof of the area formula, see for instance [1], [55], [60].

The proof of the second local variant, under the stronger assumption $\int_0^T\int_{\mathbb{R}^d}|\boldsymbol{b}_t|\,d\mu_t dt < +\infty$, is given in Proposition 8.1.8 of [12]. The same proof works under the weaker assumption (2.5).

Section 3. Many ideas of this section, and in particular the idea of looking at measures in the space of continuous maps to characterize the flow and prove its stability, are borrowed from [7], dealing with BV vector fields. Later on, the arguments have been put in a more general form, independent of the specific class of vector fields under consideration, in [9]. Here we present a more refined version of [9].

The idea of a probabilistic representation is of course classical, and appears in many contexts (particularly for equations of diffusion type); to my knowledge the first reference in the context of conservation laws and fluid mechanics is [24], where a similar approach is proposed for the incompressible Euler equation (see also [25], [26], [27]): in this case the compact (but neither metrizable, nor separable) space $X^{[0,T]}$, with $X \subset \mathbb{R}^d$ compact, has been considered.

This approach is by now a familiar one also in optimal transport theory, where transport maps and transference plans can be thought in a natural way as measures in the space of minimizing geodesics [76], and in the so called irrigation problems, a nice variant of the optimal transport problem [22]. See also [18] for a similar approach within Mather's theory. The Lecture Notes [84] (see also the Appendix of [69]) contain, among several other things, a comprehensive treatment of the topic of measures in the space of action-minimizing curves, including at the same time the optimal transport and the dynamical systems case (this unified treatment was inspired by [21]). Another related reference is [50].

The superposition principle is proved, under the weaker assumption $\int_0^T \int_{\mathbb{R}^d} |\mathbf{b}_t|^p \, d\mu_t dt < +\infty$ for some $p > 1$, in Theorem 8.2.1 of [12], see also [70] for the extension to the case $p = 1$ and to the non-homogeneous continuity equation. Very closely related results, relative to the representation of a vector field as the superposition of "elementary" vector fields associated to curves, appear in [77], [18].

In [16] an interesting variant of the stability Theorems 3.6 and 4.4 is discussed, peculiar of the case when the limit vector field $\mathbf{b}$ is a sufficiently regular gradient. In this case it has been proved in [16] that narrow convergence of μ_t^h to μ_t for all $t \in [0, T]$ and the energy estimate

$$\limsup_{h \to \infty} \int_0^T \int_{\mathbb{R}^d} |\mathbf{b}_t^h|^2 \, d\mu_t^h \, dt \le \int_0^T \int_{\mathbb{R}^d} |\mathbf{b}_t|^2 \, d\mu_t dt < +\infty$$

are sufficient to obtain the stability property. This is due to the fact that, given μ_t, gradient vector fields minimize $\int_0^T \int |c_t|^2 \, d\mu_t$ among all velocity fields c_t for which the continuity equation $\frac{d}{dt}\mu_t + D_x \cdot (c_t \mu_t) = 0$ holds (see Chapter 8 of [12] for a general proof of this fact, and for references to earlier works of Otto, Benamou-Brenier).

The convergence result in [16] can be used to answer positively a question raised in [59], concerning the convergence of the implicit Euler scheme

$$\mathbf{u}_{k+1} \in \text{Argmin}\left[\frac{1}{2h}\int_\Omega |\mathbf{u} - \mathbf{u}_k|^2 + \int_\Omega F(\nabla \mathbf{u})\, dx\right]$$

(here Ω, Ω' are bounded open in $\mathbb{R}^d$ and $\mathbf{u} : \Omega \to \Omega'$) in the case when $F(\nabla \mathbf{u})$ depends only, in a convex way, only on the determinant of $\nabla \mathbf{u}$. It turns out that, representing as in [59] $\mathbf{u}_k$ as the composition of k optimal transport maps, $\mathbf{u}_{[t/h]}$ converge as $h \downarrow 0$ to the solution $\mathbf{u}_t$ of

$$\frac{d}{dt}\boldsymbol{u}_t = \operatorname{div}\left(\nabla F(\nabla \boldsymbol{u}_t)\right),$$

built in [59] by purely differential methods (coupling a nonlinear diffusion equation for the measures $\beta_t := (\boldsymbol{u}_t)_\#(\mathcal{L}_\Omega)$ in Ω' to a transport equation for $\boldsymbol{u}_t^{-1}$). Existence of solutions (via differential or variational methods) for wider classes of energy densities F is a largely open problem.

Section 4. The definition of renormalized solution and the strong convergence of commutators are entirely borrowed from [53]. See also [54] for the relevance of this concept in connection with the existence theory for Boltzmann equation. The proof of the comparison principle assuming only an $L^1(L^1_{\mathrm{loc}})$ bound (instead of an $L^1(L^\infty)$ one, as in [53], [7]) on the divergence was suggested to me by G.Savaré. The differentiability properties of the flow have been found in [64]: later on, this differentiability property has been characterized and compared with the more classical approximate differentiability [60] in [14], while [13] contains the proof of the stronger "local" Lipschitz properties. Theorem 4.5 summarizes all these results. The paper [44] contains also more explicit Lipschitz estimates and an independent proof of the compactness of flows. See also [37] for a proof, using radial convolution kernels, of the renormalization property for vector fields satisfying $D_i \boldsymbol{b}^j + D_j \boldsymbol{b}^i \in L^1_{\mathrm{loc}}$.

Both methods, the one illustrated in these notes and the DiPerna–Lions one, are based on abstract compactness arguments and do not provide a rate of convergence in the stability theorem. It would be interesting to find an explicit rate of convergence (in mean with respect to x) of the trajectories. This problem is open even for autonomous, bounded and Sobolev (but not Lipschitz) vector fields.

No general existence result for Sobolev (or even BV) vector fields seems to be known in the infinite-dimensional case: the only reference we are aware of is [23]. Also the investigation of non-Euclidean geometries, *e.g.* Carnot groups and horizontal vector fields, could provide interesting results.

Finally, notice that the theory has a natural invariance, namely if $\boldsymbol{X}$ is a flow relative to $\boldsymbol{b}$, then $\boldsymbol{X}$ is a flow relative to $\tilde{\boldsymbol{b}}$ whenever $\{\tilde{\boldsymbol{b}} \neq \boldsymbol{b}\}$ is $\mathcal{L}^{1+d}$-negligible in $(0,T) \times \mathbb{R}^d$. So a natural question is whether the uniqueness "in the selection sense" might be enforced by choosing a canonical representative $\tilde{\boldsymbol{b}}$ in the equivalence class of $\boldsymbol{b}$: in other words we may think that, for a suitable choice of $\tilde{\boldsymbol{b}}$, the ODE $\dot{\gamma}(t) = \tilde{\boldsymbol{b}}_t(\gamma(t))$ has a unique absolutely continuous solution starting from x for $\mathcal{L}^d$-a.e. x.

Section 5. Here we followed closely [7]. The main idea of this section, *i.e.* the adaptation of the convolution kernel to the local behaviour of the vector field, has been used at various level of generality in [30], [66], [41] (see also [38], [39] for related results independent of this technique), until the general result [7].

The optimal regularity condition on $\boldsymbol{b}$ ensuring the renormalization property, and therefore the validity of the comparison principle in $\mathcal{L}_b$, is still not known. New results, both in the Sobolev and in the BV framework, are presented in [11], [64], [65].

In [15] we investigate in particular the possibility to prove the renormalization property for nearly incompressible $BV_{loc} \cap L^\infty$ fields $\boldsymbol{b}$: they are defined by the property that there exists a positive function ρ, with $\ln \rho \in L^\infty$, such that the space-time field $(\rho, \rho\boldsymbol{b})$ is divergence free. As in the case of the Keyfitz–Kranzer system, the existence a function ρ with this property seems to be a natural replacement of the condition $D_x \cdot \boldsymbol{b} \in L^\infty$ (and is actually implied by it); as explained in [10], a proof of the renormalization property in this context would lead to a proof of a conjecture, due to Bressan, on the compactness of flows associated to a sequence of vector fields bounded in $BV_{t,x}$.

Section 6. In connection with the Keyfitz–Kranzer system there are several open questions: in particular one would like to obtain uniqueness (and stability) of the solution in more general classes of admissible functions (partial results in this direction are given in [10]). A strictly related problem is the convergence of the vanishing viscosity method to the solution built in [8]. Also, very little about the regularity of solutions is presently known: we know [49] that BV estimates do not hold and, besides, that the contruction in [8] seems not applicable to more general systems of triangular type, see the counterexample in [43].

In connection with the semi-geostrophic problem, the main problem is the existence of solutions in the physical variables, i.e. in the Eulerian form. A formal argument suggests that, given P_t, the velocity $\boldsymbol{u}$ should be defined by

$$\partial_t \nabla P_t^*(\nabla P_t(x)) + \nabla^2 P_t^*(\nabla P_t(x))J(\nabla P_t(x) - x).$$

On the other hand, the a-priori regularity on ∇P_t (ensured by the convexity of P_t) is a BV regularity, and it is still not clear how this formula could be rigorously justified. In this connection, an important intermediate step could be the proof of the $W^{1,1}$ regularity of the maps ∇P_t (see also [33], [34], [35], [36], [80], [81] for the regularity theory of optimal transport maps under regularity assumptions on the initial and final densities).

References

1. M. AIZENMAN: *On vector fields as generators of flows: a counterexample to Nelson's conjecture.* Ann. Math., **107** (1978), 287–296.
2. G. ALBERTI: *Rank-one properties for derivatives of functions with bounded variation.* Proc. Roy. Soc. Edinburgh Sect. A, **123** (1993), 239–274.
3. G. ALBERTI & L. AMBROSIO: *A geometric approach to monotone functions in* $\mathbb{R}^n$. Math. Z., **230** (1999), 259–316.
4. G. ALBERTI & S. MÜLLER: *A new approach to variational problems with multiple scales.* Comm. Pure Appl. Math., **54** (2001), 761–825.
5. F.J. ALMGREN: *The theory of varifolds – A variational calculus in the large,* Princeton University Press, 1972.
6. L. AMBROSIO, N. FUSCO & D. PALLARA: *Functions of bounded variation and free discontinuity problems.* Oxford Mathematical Monographs, 2000.

7. L. AMBROSIO: *Transport equation and Cauchy problem for BV vector fields.* Inventiones Mathematicae, **158** (2004), 227–260.

8. L. AMBROSIO & C. DE LELLIS: *Existence of solutions for a class of hyperbolic systems of conservation laws in several space dimensions.* International Mathematical Research Notices, **41** (2003), 2205–2220.

9. L. AMBROSIO: *Lecture notes on transport equation and Cauchy problem for BV vector fields and applications.* Preprint, 2004 (available at http://cvgmt.sns.it).

10. L. AMBROSIO, F. BOUCHUT & C. DE LELLIS: *Well-posedness for a class of hyperbolic systems of conservation laws in several space dimensions.* Comm. PDE, **29** (2004), 1635–1651.

11. L. AMBROSIO, G. CRIPPA & S. MANIGLIA: *Traces and fine properties of a BD class of vector fields and applications.* Preprint, 2004 (to appear on Annales de Toulouse).

12. L. AMBROSIO, N. GIGLI & G. SAVARÉ: *Gradient flows in metric spaces and in the Wasserstein space of probability measures.* Lectures in Mathematics, ETH Zurich, Birkhäuser, 2005.

13. L. AMBROSIO, M. LECUMBERRY & S. MANIGLIA: *Lipschitz regularity and approximate differentiability of the DiPerna-Lions flow.* Preprint, 2005 (available at http://cvgmt.sns.it and to appear on Rend. Sem. Fis. Mat. di Padova).

14. L. AMBROSIO & J. MALÝ: *Very weak notions of differentiability.* Preprint, 2005 (available at http://cvgmt.sns.it).

15. L. AMBROSIO, C. DE LELLIS & J. MALÝ: *On the chain rule for the divergence of BV like vector fields: applications, partial results, open problems.* Preprint, 2005 (available at http://cvgmt.sns.it).

16. L. AMBROSIO, S. LISINI & G. SAVARÉ: *Stability of flows associated to gradient vector fields and convergence of iterated transport maps.* In preparation.

17. E.J. BALDER: *New fundamentals of Young measure convergence.* CRC Res. Notes in Math. **411**, 2001.

18. V. BANGERT: *Minimal measures and minimizing closed normal one-currents.* Geom. funct. anal., **9** (1999), 413–427.

19. J. BALL & R. JAMES: *Fine phase mixtures as minimizers of energy.* Arch. Rat. Mech. Anal., **100** (1987), 13–52.

20. J.-D. BENAMOU & Y. BRENIER: *Weak solutions for the semigeostrophic equation formulated as a couples Monge-Ampere transport problem.* SIAM J. Appl. Math., **58** (1998), 1450–1461.

21. P. BERNARD & B. BUFFONI: *Optimal mass transportation and Mather theory.* Preprint, 2004.

22. M. BERNOT, V. CASELLES & J.M. MOREL: *Traffic plans.* Preprint, 2004.

23. V. BOGACHEV & E.M. WOLF: *Absolutely continuous flows generated by Sobolev class vector fields in finite and infinite dimensions.* J. Funct. Anal., **167** (1999), 1–68.

24. Y. BRENIER: *The least action principle and the related concept of generalized flows for incompressible perfect fluids.* J. Amer. Mat. Soc., **2** (1989), 225–255.

25. Y. BRENIER: *The dual least action problem for an ideal, incompressible fluid.* Arch. Rational Mech. Anal., **122** (1993), 323–351.

26. Y. BRENIER: *A homogenized model for vortex sheets.* Arch. Rational Mech. Anal., **138** (1997), 319–353.

27. Y. BRENIER: *Minimal geodesics on groups of volume-preserving maps and generalized solutions of the Euler equations.* Comm. Pure Appl. Math., **52** (1999), 411–452.

28. F. BOUCHUT & F. JAMES: *One dimensional transport equation with discontinuous coefficients.* Nonlinear Analysis, **32** (1998), 891–933.

29. F. BOUCHUT, F. GOLSE & M. PULVIRENTI: *Kinetic equations and asymptotic theory.* Series in Appl. Math., Gauthiers-Villars, 2000.

30. F. BOUCHUT: *Renormalized solutions to the Vlasov equation with coefficients of bounded variation.* Arch. Rational Mech. Anal., **157** (2001), 75–90.

31. F. BOUCHUT, F. JAMES & S. MANCINI: *Uniqueness and weak stability for multidimensional transport equations with one-sided Lipschitz coefficients.* Preprint, 2004 (to appear on Annali Scuola Normale Superiore).

32. A. BRESSAN: *An ill posed Cauchy problem for a hyperbolic system in two space dimensions.* Rend. Sem. Mat. Univ. Padova, **110** (2003), 103–117.

33. L.A. CAFFARELLI: *Some regularity properties of solutions of Monge Ampère equation,* Comm. Pure Appl. Math., **44** (1991), 965–969.

34. L.A. CAFFARELLI: *Boundary regularity of maps with convex potentials,* Comm. Pure Appl. Math., **45** (1992), 1141–1151.

35. L.A. CAFFARELLI: *The regularity of mappings with a convex potential.* J. Amer. Math. Soc., **5** (1992), 99–104.

36. L.A. CAFFARELLI: *Boundary regularity of maps with convex potentials.*, Ann. of Math., **144** (1996), 453–496.

37. I. CAPUZZO DOLCETTA & B. PERTHAME: *On some analogy between different approaches to first order PDE's with nonsmooth coefficients.* Adv. Math. Sci Appl., **6** (1996), 689–703.

38. A. CELLINA: *On uniqueness almost everywhere for monotonic differential inclusions.* Nonlinear Analysis, TMA, **25** (1995), 899–903.

39. A. CELLINA & M. VORNICESCU: *On gradient flows.* Journal of Differential Equations, **145** (1998), 489–501.

40. F. COLOMBINI & N. LERNER: *Uniqueness of continuous solutions for BV vector fields.* Duke Math. J., **111** (2002), 357–384.

41. F. COLOMBINI & N. LERNER: *Uniqueness of L^∞ solutions for a class of conormal BV vector fields.* Preprint, 2003.

42. F. COLOMBINI, T. LUO & J. RAUCH: *Uniqueness and nonuniqueness for nonsmooth divergence-free transport.* Preprint, 2003.

43. G. CRIPPA & C. DE LELLIS: *Oscillatory solutions to transport equations.* Preprint, 2005 (available at http://cvgmt.sns.it).

44. G. CRIPPA & C. DE LELLIS: *Estimates for transport equations and regularity of the DiPerna-Lions flow.* In preparation.

45. M. CULLEN: *On the accuracy of the semi-geostrophic approximation.* Quart. J. Roy. Metereol. Soc., **126** (2000), 1099–1115.

46. M. CULLEN & W. GANGBO: *A variational approach for the 2-dimensional semigeostrophic shallow water equations.* Arch. Rational Mech. Anal., **156** (2001), 241–273.

47. M. CULLEN & M. FELDMAN: *Lagrangian solutions of semigeostrophic equations in physical space.* Preprint, 2003.

48. C. DAFERMOS: *Hyperbolic conservation laws in continuum physics.* Springer Verlag, 2000.

49. C. DE LELLIS: *Blow-up of the BV norm in the multidimensional Keyfitz and Kranzer system.* Duke Math. J., **127** (2004), 313–339.

50. L. DE PASCALE, M.S. GELLI & L. GRANIERI: *Minimal measures, one-dimensional currents and the Monge-Kantorovich problem.* Preprint, 2004 (available at http://cvgmt.sns.it).

51. N. DE PAUW: *Non unicité des solutions bornées pour un champ de vecteurs BV en dehors d'un hyperplan.* C.R. Math. Sci. Acad. Paris, **337** (2003), 249–252.

52. R.J. DIPERNA: *Measure-valued solutions to conservation laws.* Arch. Rational Mech. Anal., **88** (1985), 223–270.

53. R.J. DI PERNA & P.L. LIONS: *Ordinary differential equations, transport theory and Sobolev spaces.* Invent. Math., **98** (1989), 511–547.

54. R.J. DI PERNA & P.L. LIONS: *On the Cauchy problem for the Boltzmann equation: global existence and weak stability.* Ann. of Math., **130** (1989), 312–366.

55. L.C. EVANS & R.F. GARIEPY: *Lecture notes on measure theory and fine properties of functions,* CRC Press, 1992.

56. L.C. EVANS: *Partial Differential Equations.* Graduate studies in Mathematics, **19** (1998), American Mathematical Society.

57. L.C. EVANS: *Partial Differential Equations and Monge–Kantorovich Mass Transfer.* Current Developments in Mathematics, 1997, 65–126.

58. L.C. EVANS & W. GANGBO: *Differential equations methods for the Monge-Kantorovich mass transfer problem.* Memoirs AMS, **653**, 1999.

59. L.C. EVANS, W. GANGBO & O. SAVIN: *Nonlinear heat flows and diffeomorphisms.* Preprint, 2004.

60. H. FEDERER: *Geometric measure theory,* Springer, 1969.

61. M. HAURAY: *On Liouville transport equation with potential in BV_{loc}.* Comm. in PDE, **29** (2004), 207–217.

62. M. HAURAY: *On two-dimensional Hamiltonian transport equations with L^p_{loc} coefficients.* Ann. IHP Nonlinear Anal. Non Linéaire, **20** (2003), 625–644.

63. B.L. KEYFITZ & H.C. KRANZER: *A system of nonstrictly hyperbolic conservation laws arising in elasticity theory.* Arch. Rational Mech. Anal. **1980**, 72, 219–241.

64. C. LE BRIS & P.L. LIONS: *Renormalized solutions of some transport equations with partially $W^{1,1}$ velocities and applications.* Annali di Matematica, **183** (2003), 97–130.

65. N. LERNER: *Transport equations with partially BV velocities.* Preprint, 2004.

66. P.L. LIONS: *Sur les équations différentielles ordinaires et les équations de transport.* C. R. Acad. Sci. Paris Sér. I, **326** (1998), 833–838.

67. P.L. LIONS: *Mathematical topics in fluid mechanics, Vol. I: incompressible models.* Oxford Lecture Series in Mathematics and its applications, **3** (1996), Oxford University Press.

68. P.L. LIONS: *Mathematical topics in fluid mechanics, Vol. II: compressible models.* Oxford Lecture Series in Mathematics and its applications, **10** (1998), Oxford University Press.

69. J. LOTT & C. VILLANI: *Weak curvature conditions and Poincaré inequalities.* Preprint, 2005.

70. S. MANIGLIA: *Probabilistic representation and uniqueness results for measure-valued solutions of transport equations.* Preprint, 2005.

71. J.N. MATHER: *Minimal measures.* Comment. Math. Helv., **64** (1989), 375–394.

72. J.N. MATHER: *Action minimizing invariant measures for positive definite Lagrangian systems.* Math. Z., **207** (1991), 169–207.

73. E.Y. PANOV: *On strong precompactness of bounded sets of measure-valued solutions of a first order quasilinear equation.* Math. Sb., **186** (1995), 729–740.

74. G. PETROVA & B. POPOV: *Linear transport equation with discontinuous coefficients.* Comm. PDE, **24** (1999), 1849–1873.

75. F. POUPAUD & M. RASCLE: *Measure solutions to the liner multidimensional transport equation with non-smooth coefficients.* Comm. PDE, **22** (1997), 337–358.

76. A. PRATELLI: *Equivalence between some definitions for the optimal transport problem and for the transport density on manifolds.* preprint, 2003, to appear on Ann. Mat. Pura Appl (available at http://cvgmt.sns.it).

77. S.K. SMIRNOV: *Decomposition of solenoidal vector charges into elementary solenoids and the structure of normal one-dimensional currents.* St. Petersburg Math. J., **5** (1994), 841–867.

78. L. TARTAR: *Compensated compactness and applications to partial differential equations.* Research Notes in Mathematics, Nonlinear Analysis and Mechanics, ed. R. J. Knops, vol. **4**, Pitman Press, New York, 1979, 136–211.

79. R. TEMAM: *Problémes mathématiques en plasticité.* Gauthier-Villars, Paris, 1983.

80. J.I.E. URBAS: *Global Hölder estimates for equations of Monge-Ampère type,* Invent. Math., **91** (1988), 1–29.

81. J.I.E. URBAS: *Regularity of generalized solutions of Monge-Ampère equations,* Math. Z., **197** (1988), 365–393.

82. A. VASSEUR: *Strong traces for solutions of multidimensional scalar conservation laws.* Arch. Ration. Mech. Anal., **160** (2001), 181–193.

83. C. VILLANI: *Topics in mass transportation.* Graduate Studies in Mathematics, **58** (2004), American Mathematical Society.

84. C. VILLANI: *Optimal transport: old and new.* Lecture Notes of the 2005 Saint-Flour Summer school.

85. L.C. YOUNG: *Lectures on the calculus of variations and optimal control theory,* Saunders, 1969.

Issues in Homogenization for Problems with Non Divergence Structure

Luis Caffarelli[1] and Luis Silvestre[2]

[1] Department of Mathematics, University of Texas at Austin
   1 University Station C1200 Austin, TX 78712-0257, USA
   `caffarel@math.utexas.edu`
[2] Department of Mathematics,University of Texas at Austin
   `lsilvest@math.utexas.edu`

1 Introduction

When we look at a differential equation in a very irregular media (composite material, mixed solutions, etc.) from very close, we may see a very complicated problem. However, if we look from far away we may not see the details and the problem may look simpler. The study of this effect in partial differential equations is known as homogenization. The effect of the inhomogeneities oscillating at small scales is often not a simple average and may be hard to predict: a geodesic in an irregular medium will try to avoid the bad areas, the roughness of a surface may affect in nontrivial way the shapes of drops laying on it, etc...

The purpose of these notes is to discuss three problems in homogenization and their interplay.

In the first problem, we consider the homogenization of a free boundary problem. We study the shape of a drop lying on a rough surface. We discuss in what case the homogenization limit converges to a perfectly round drop. It is taken mostly from the joint work with Antoine Mellet [5].

The second problem concerns the construction of plane like solutions to the minimal surface equation in periodic media. This is related to homogenization of minimal surfaces. The details can be found in the joint paper with Rafael de la Llave [2].

The third problem concerns existence of homogenization limits for solutions to fully nonlinear equations in ergodic random media. It is mainly based on the joint paper with Panagiotis Souganidis and Lihe Wang [7].

We will try to point out the main techniques and the common aspects. The focus has been set to the basic ideas. The main purpose is to make this advanced topics as readable as possible. In every case, the original papers are referenced.

2 Homogenization of a Free Boundary Problem: Capillary Drops

The shape of a drop lying on a surface tries to minimize its energy for a given volume. The energy has a term proportional to the capillary area A between the water and the air, another term related to the contact area W between the drop and the surface, and a third term related to the gravitational potential energy.

$$\text{Energy} = \sigma A - \sigma \beta W + \Gamma \text{ Gravitational Energy}$$

For the time being, we will neglect the effect of gravity ($\Gamma = 0$) and consider $\sigma = 1$.

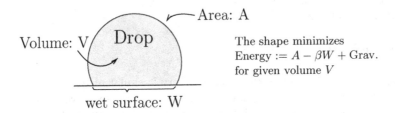

Volume: V   Drop   Area: A

The shape minimizes
Energy $:= A - \beta W + \text{Grav.}$
for given volume V

wet surface: W

Fig. 1. A drop lying on a plane surface

The surface of the drop that is not in contact with the floor will have a constant mean curvature. We can see this perturbing its shape in a way that we preserve volume. If we add a bit of volume around a point and we subtract the same amount around another point, we obtain another admissible shape and so the corresponding area must increase. This implies that the mean curvature at both points must coincide.

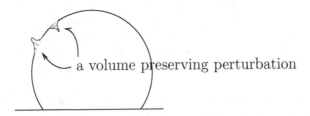

a volume preserving perturbation

Fig. 2. Suitable perturbations show that the free surface has a constant mean curvature

The parameter β is a real number between -1 and 1 that depends on the surface and is the relative adhesion coefficient between the fluid and the surface. Its effect on the shape of the drop is to prescribe the contact angle at the free boundary: $\cos\gamma = \beta$.

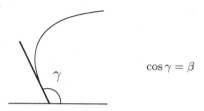

Fig. 3. The contact angle depends on β

A value $\beta > 0$ will cause the shape of the drop to expand trying to span a larger wet surface. When $\beta < 0$ (hydrophobic surface) on the other hand, the wet surface will tend to shrink. In the limit case $\beta = 1$ the wet surface would try to cover the whole plane, whereas for $\beta = -1$, the optimal shape would be a sphere that does not touch the floor at all.

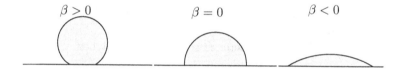

Fig. 4. Different shapes depending on the value of β

Under these conditions, it can be shown that there is a minimizer for the energy, and the shape of the corresponding drop is given by a sphere cap. The case we are interested however is when the drop rests on an irregular surface. Namely, we will consider a variable $\beta(x)$, oscillating fast and bounded so that $|\beta(x)| \leq \lambda < 1$. To capture the effect of a very oscillating adhesion coefficient, we fix a periodic function β and consider $\beta(x/\varepsilon)$ for a small ε. The energy is then given by

$$J_\varepsilon = A - \int_{\text{wet surface}} \beta\left(\frac{x}{\varepsilon}\right) \, dx \tag{1}$$

Our purpose is to study the existence and regularity for a given $\varepsilon > 0$ of a shape that minimizes the energy. And we want to understand the way it behaves as $\varepsilon \to 0$. We will see that the absolute minimizers of J_ε converge uniformly to a spherical cap that corresponds to the minimizer of

$$J_0 = A - \langle \beta \rangle W$$

where $\langle \beta \rangle = f\,\beta\,dx$ is the average of β.

However, the same conclusion cannot be taken for other critical points of J_ε. In general, the shapes of drops will not achieve an absolute minimizer. Any local minimum of J_ε would be a stable shape for a drop. The limits of these other solutions behave in a way that is harder to predict and many interesting phenomena can be observed. The most spectacular effect is the hysteresis: the contact angle depends on how the drop was formed. If the drop was formed by advancing the liquid, the final contact angle is greater than $\langle \beta \rangle$. If the equilibrium was achieved by receding the liquid (like in a process of evaporation), the angle obtained is less than $\langle \beta \rangle$.

The existence of a minimizer for each ε can be done in a very classical way in the framework of sets of finite perimeter. We will study some regularity properties of the minimizers. First we will show that the surface of the drop separates volume in a more or less balanced way. Secondly, we will see that the boundary of the contact set has a finite $n - 1$ Hausdorff measure. Then we will use those estimates together with a stability result to show that the minimizers of J_ε converge to spherical caps as $\varepsilon \to 0$. To conclude this part, we will discuss the phenomena of Hysteresis.

2.1 Existence of a Minimizer

In order to prove existence, we have to work in the framework of boundaries of sets of finite perimeter.

Roughly, a set of finite perimeter Ω is the limit of polyhedra, Ω_k, of finite area, i.e.

$$|\Omega \Delta \Omega_k| \to 0$$

and $Area(\partial \Omega_k) \leq C$ for all k.

Sets of finite perimeter are defined up to sets of measure zero. We normalize E so that

$$0 < \left| \bar{E} \cap B_r(x) \right| < |B_r(x)| \quad \text{for all } x \in E \text{ and } r > 0$$

There is a well established theory for such sets. The classical reference is [13].

We will consider a set $E \subset \mathbb{R}^n \times [0, +\infty)$ that represents the shape of the drop. We denote (x, z) an arbitrary point with $x \in \mathbb{R}^n$ and $z \in [0, +\infty)$. Our energy functional reads

$$J_\varepsilon(E) = Area(\partial E \cap \{z > 0\}) - \int_{z=0} \beta \left(\frac{x}{\varepsilon} \right) \chi_E \, dx \tag{2}$$

(In the following, we will omit the ε in J_ε unless it is necessary to stress it out).

The theory of finite perimeter set provides the necessary compactness results to show existence of a minimizer, as long as we restrict E to be a subset of a bounded set $\Gamma_{RT} := \{|x| < R, z < T\}$. Of course, we must take R and

T large enough so that we can fit at least one set E of volume V inside. To obtain an unrestricted minimizer of (2), we must prove that for R and T large enough, there is one corresponding minimizer E_{RT} that does not touch the boundary of Γ_{RT}. Since β is periodic, it is enough to show that E_{RT} remains bounded independently of R and T. If the diameter of E_{RT} is less than $R/2$, we can translate it an integer multiple of ε inside of Γ_{RT} to obtain an unrestricted minimizer. The detailed proof can be found in [5]

2.2 Positive Density Lemmas

The first regularity results we obtain for minimizers are related to the nondegenerate way the surface of the drop separates volume. All the proofs of these lemmas follow the same idea. An ordinary differential equation is constructed that exploits the nonlinearity of the isoperimetric inequality.

But before, we will make a few simple observations. Let E be a minimizer for a volume V_0, and let A be its free perimeter $(A = Area(\partial E \cap \{z > 0\}))$. Above every point on the wet surface $E \cap \{z = 0\}$, there must be a point in the free surface: ∂E. Then

$$A \geq \int_{z=0} \chi_E \, dx \geq \frac{1}{\lambda} \left| \int_{z=0} \beta \left(\frac{x}{\varepsilon} \right) \, dx \right|$$

And therefore

$$(1 - \lambda)A \leq J(E)$$

From the isoperimetric inequality we have $A \geq w_{n+1} V_0^{\frac{n}{n+1}}$. Since a sphere B with volume V_0 that does not touch the floor $\{z = 0\}$ is an admissible set, we also have:

$$(1 - \lambda)A \leq J(E) \leq J(B) = w_{n+1} V_0^{\frac{n}{n+1}}$$

And thus we have both estimates:

$$c_0 V_0^{\frac{n}{n+1}} \leq A \leq C_1 V_0^{\frac{n}{n+1}}$$

Now we want to compare the minimum energy for two different volumes.

$$\min_{volume=V_0} J \leq \min_{volume=V_0+\delta} J \leq \min_{volume=V_0} J + C_1 V_0^{-\frac{1}{n+1}} \delta \qquad (3)$$

The first inequality can be obtained simply taking the minimizer for $volume = V_0 + \delta$ and chopping a piece at the top of volume δ. Thus we obtain an admissible set of volume V_0 for which the energy J decreased.

For the second inequality, we consider the set E with volume V_0 that minimizes J and take a vertical dilation $E_t = \{(x, t) : (x, (1 + t)^{-1} z) \in E\}$. Then for $t = \delta/V_0$, E_t is an admissible set of volume $V_0 + \delta$. The contact surface did not change, so its only difference in the energy is given by the free surface. Let A be the free perimeter of E, then the perimeter of E_t is less than $(1 + t)A$, so their respective energies differ at most by tA. Then

$$\min_{volume=V_0+\delta} J - \min_{volume=V_0} J \leq tA$$

$$\leq \frac{\delta}{V_0} c_1 V_0^{\frac{n}{n+1}}$$

$$\leq c_1 V_0^{-\frac{1}{n+1}} \delta$$

The first lemma we want to prove is actually a classical result in minimal surfaces adapted to this case. We will come back to this lemma again when we study plane like minimal surfaces in periodic media in the second part of these notes.

Before starting with the lemmas it is worth to point out an elementary fact of calculus that will come handy. If we have a nonnegative function u such that $u' \geq cu^{\frac{n+1}{n}}$ then u is a nondecreasing function that can stay equal to zero for any amount of time. But if $t_0 = \sup\{t : u(t) = 0\}$, then $u(t) \geq c(t - t_0)^n$ for any $t > t_0$.

Lemma 2.1. Let $(x_0, z_0) \in \partial E$ with $z_0 > 0$. There exists a universal constant c such that for all $r < z_0$ we have

$$|B_r(x_0, z_0) \cap E| \geq cr^{n+1}$$
$$|B_r(x_0, z_0) \setminus E| \geq cr^{n+1}$$

Proof. We define

$$U_1(r) = |B_r(x_0, z_0) \setminus E| \qquad S_1(r) = Area(\partial B_r(x_0, z_0) \setminus E)$$
$$U_2(r) = |B_r(x_0, z_0) \cap E| \qquad S_2(r) = Area(\partial B_r(x_0, z_0) \cap E)$$
$$A(r) = Area(B_r \cap \partial E)$$

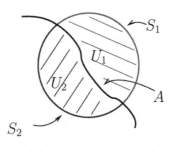

By estimating $J(E \cup B_r)$ and $J(E \setminus B_r)$ and using (3), we can compare S_1 and S_2 to A.

$$J(E \cup B_r) \geq \min_{volume=V_0+U_1} J \geq J(E)$$

$$J(E) + S_1 - A \geq J(E)$$

$$S_1 - A \geq 0$$

We also know by the isoperimetrical inequality that $U_1^{\frac{n}{n+1}} \leq C(A + S_1)$. If we combine this with the above inequality we obtain

$$U_1^{\frac{n}{n+1}} \leq CS_1$$

But now we observe that $S_1(r) = U_1'(r)$, so we obtain the ODE: $U_1'(r) \geq cU_1^{\frac{n}{n+1}}$. Moreover, we know $U_1(0) = 0$ and $U_1(r) > 0$ for any $r > 0$. This implies the result of the lemma.

For U_2, a similar argument is done using the other inequality in (3).

With almost the same proof, we can also obtain a similar lemma for (x_0, z_0) in the boundary of the wet surface $E \cap \{z = 0\}$.

Lemma 2.2. *Given $x_0 \in \mathbb{R}^n$, let $\Gamma_{rt} = \{(x, z) : |x - x_0| \leq r \wedge 0 \leq z \leq t\}$.*
There exist two universal constants $c_0, c_1 > 0$ such that for any minimizer E of J with volume V_0 such that

$$\{(x, t) : |x - x_0| \leq r_0\} \subset E \quad (\text{resp. } \subset CE)$$
$$\exists z \in (0, t) \text{ such that } (x, z) \in \partial E$$

then

$$|CE \cap \Gamma_{rt}| \geq c_0 r^{n+1} \quad (\text{resp. } |E \cap \Gamma_{rt}| \geq c_0 r^{n+1})$$

for all $r < r_0$ (resp. for all $r < r_0$ such that $|E \cap \Gamma_{rt}| \leq c_1 V_0$).

Remark. When we say $\{(x, t) : |x - x_0| \leq r\} \subset E$, we actually mean that the trace of E on $\{(x, t) : |x - x_0| \leq r\}$ is constant 1. Sets of finite perimeter have a well defined trace in L^1.

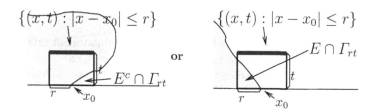

Fig. 5. Lemma 2.2

Proof. We proceed in a similar fashion as in the proof of Lemma 2.1. Let

$$U(r) = |\Gamma_{rt} \setminus E|$$
$$S(r) = Area(\partial \Gamma_{rt} \setminus E)$$
$$A(r) = Area(\Gamma_{rt} \cap \partial E)$$
$$W(r) = Area(\{z = 0\} \cap \Gamma_{rt} \setminus E) = \int_{z=0 \wedge |x-x_0|<r} (1 - \chi_E) \, dx$$

$$w(r) = \int_{z=0 \wedge |x-x_0|<r} \beta\left(\frac{x}{\varepsilon}\right)(1 - \chi_E)\,dx$$

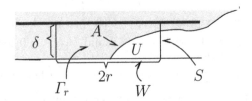

Since above any point in the wet surface, there is a point in $\partial E \cap \Gamma_{rt}$, $W \le A$. Therefore, also $|w| \le \lambda A$.

By comparing $J(E)$ with $J(E \cup \Gamma_{rt})$, we get

$$J(E) \le J(E \cup \Gamma_r)$$
$$J(E) \le J(E) + S(r) - A(r) + w(r)$$
$$0 \le S(r) - A(r) + w(r)$$
$$0 \le S(r) - (1-\lambda)A(r)$$

By the isoperimetric inequality we know that

$$U^{\frac{n}{n+1}} \le c(A + S + W)$$

Combining the above inequalities we obtain:

$$U^{\frac{n}{n+1}} \le C\,S(r)$$

And we observe that $S(r) = U'(r)$ to obtain the nonlinear ODE: $U'(r) \ge cU^{\frac{n}{n+1}}$. Moreover, $U(0) = 0$ and $U(r) > 0$ for any $r > 0$, then $U(r) > cr^{n+1}$.

This proves the first case of the lemma. The other case follows almost in the same way but exchanging E and CE. Since in that case we have to use the other inequality in (3), we must use that $|R \cap \Gamma_{rt}| \le c_1 V_0$ to control the extra term.

Corollary 2.1. *If $(x_0, 0) \in \partial E$, then*

$$\left| E \cap B_r^+(x_0, 0) \right| \ge c\,r^{n+1}$$
$$\left| CE \cap B_r^+(x_0, 0) \right| \ge c\,r^{n+1}$$

for every r such that $|E \cap B_r^+(x_0, 0)| \le c_1 V_0$.

Proof. The set $B_{r/2}^+(x_0, 0) \setminus \{z < \delta_0 r/2\}$ is either completely contained in E or CE, or the set $B_{r/2}^+(x_0, 0) \setminus \{z < \delta_0 r/2\} \cap \partial E$ is not empty.

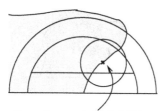 

Either there is a point of ∂E here   or this region.is completely contained in either E or E^c

In the first case, we apply Lemma 2.2 to obtain that both

$$|B_r^+(x_0,0) \cap E| \geq cr^{n+1}$$
$$|B_r^+(x_0,0) \setminus E| \geq cr^{n+1}$$

In the second case, there is a $(x_0, z_0) \in B_{r/2}^+(x_0,0) \setminus \{z < \delta_0 r/2\} \cap \partial E$, then we use 2.1 for a ball centered at (x_0, z_0) with radius $r/4$ to obtain also

$$|B_r^+(x_0,0) \cap E| \geq cr^{n+1}$$
$$|B_r^+(x_0,0) \setminus E| \geq cr^{n+1}$$

Corollary 2.2. *If* $(x_0,0) \in \partial E$, *then*

$$Area(\partial E \cap B_r^+(x_0,0)) \geq c\, r^n$$

for every r *such that* $|E \cap B_r^+(x_0,0)| \leq c_1 V_0$.

Proof. This is a consequence of Corollary 2.1 combined with the isoperimetric inequality.

2.3 Measure of the Free Boundary

Our goal now is to show that the boundary of the wet surface $\partial(E \cap \{z = 0\})$ in $\mathbb{R}^n$ has a finite $n - 1$ Hausdorff measure. We will do it by estimating the area of the drop close to it.

Now we will estimate the area of the drop that is close to the floor, and then we will obtain an estimate on the $n - 1$ Hausdorff measure of the free boundary by a covering argument using the previous lemma.

Lemma 2.3. *There exists a constant* C *such that*

$$Area(\partial E \cap \{0 < z < t\}) \leq CV^{\frac{n-1}{n+1}}t$$

Proof. We will cut from E all the points for which $z < t$ and lower it to touch the floor again. We call F the set that we obtain (i.e. $F = \{(x, z) : (x, z+t) \in E\}$).

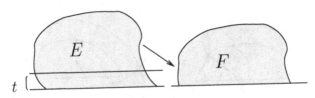

Since E is bounded, $|F| \leq |E| - Ct$, and thanks to (3) we have $J(E) \leq J(F) + Ct$. Moreover

$$J(E) - J(F) = Area(\partial E \cap \{0 < z < t\}) - \int_{z=0} \beta \left(\frac{x}{\varepsilon}\right) (\chi_E(x,0) - \chi_F(x,0)) \, dx$$

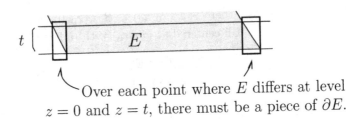

Over each point where E differs at level $z = 0$ and $z = t$, there must be a piece of ∂E.

But if x belongs to the difference between $E \cap \{z = 0\}$ and $E \cap \{z = t\}$, then there must be a $z \in (0, t)$ such that $(x, z) \in \partial E$. Therefore

$$\int_{z=0} |\chi_E(x,0) - \chi_F(x,0)| \, dx \leq Area(\partial E \cap \{0 < z < t\})$$

Thus we obtain

$$\min(1, 1 - \lambda) Area(\partial E \cap \{0 < z < t\}) \leq CV^{\frac{n-1}{n+1}} t$$

which concludes the proof.

We are now ready to establish the $n - 1$ Hausdorff estimate on the free boundary.

Theorem 2.1. *The contact line $\partial(E \cap \{z = 0\})$ in $\mathbb{R}^n$ has finite $n - 1$ Hausdorff measure and*

$$\mathcal{H}^{\frac{n-1}{n+1}}(\partial(E \cap \{z = 0\})) \leq CV^{\frac{n-1}{n+1}}$$

Proof. We consider a covering of $\partial(E \cap \{z = 0\})$ with balls of radius r and finite overlapping.

From Lemma 2.2, in each ball there is at least cr^n area. But by Lemma 2.3, the total area does not exceed $CV^{\frac{n-1}{n+1}} r$. Thus, the number of balls cannot exceed $CV^{\frac{n-1}{n+1}} r^{-(n-1)}$. Which proves the result.

2.4 Limit as $\varepsilon \to 0$

The $n-1$ Hausdorff estimate of the free boundary will help us prove that the minimizers E converge uniformly to a spherical cap as $\varepsilon \to 0$.

Let $\langle \beta \rangle$ be the average of β in the unit cube: $\langle \beta \rangle = \int_{Q_1} \beta \, dx$ and

$$J_0(E) = Area(\partial E \cap \{z > 0\}) + \langle \beta \rangle Area(E \cap \{z = 0\}) \qquad (4)$$

As it was mentioned before, the minimizer of J_0 from all the sets with a given volume V is a spherical cap B_ρ^+ such that $|B_\rho^+| = V$ and the cosine of its contact angle is $\langle \beta \rangle$.

Let us check how different $J(E)$ and $J_0(E)$ are. Their only difference is in the term related to the wet surface. Recall that $\beta \left(\frac{x}{\varepsilon}\right)$ is periodic in cubes of size ε. For every such cube that is completely contained inside the wet surface of E it is the same to integrate $\beta \left(\frac{x}{\varepsilon}\right)$ or to integrate the average of β. The difference of $J(E)$ and $J_0(E)$ is then given only by the cells that intersect the boundary of $(E \cap \{z = 0\})$.

But according the the $n-1$ Hausdorff estimate of the free boundary, the number of such cells cannot exceed $CV^{\frac{n-1}{n+1}} \varepsilon^{1-n}$. Since the volume of each cell is ε^n we deduce:

$$|J_0(E) - J(E)| \le C\lambda V^{\frac{n-1}{n+1}} \varepsilon$$

The same conclusion can be taken for B_ρ^+:

$$\left|J_0(B_\rho^+) - J(B_\rho^+)\right| \le C\lambda V^{\frac{n-1}{n+1}} \varepsilon$$

And noticing that $J(E) \le J(B_\rho^+)$ and $J_0(B_\rho^+) \le J_0(E)$ we obtain

$$\left|J_0(E) - J_0(B_\rho^+)\right| \le C\lambda V^{\frac{n-1}{n+1}} \varepsilon$$

The convergence of E to B_ρ^+ is then a consequence of the following stability theorem whose proof we omit.

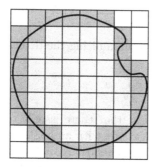

Fig. 6. In the inner cubes, it is the same to integrate $\beta(x/\varepsilon)$ or its average. The difference between J_0 and J is concentrated in the cells that intersect the free boundary

Theorem 2.2. *Let $E \subset B_R \times [0, R)$ such that $J_0(E) \leq J_0(B_\rho^+) + \delta$. Then there exists a universal $\alpha > 0$ and a constant C (depending on R) such that*

$$\left| E \triangle B_\rho^+ \right| \leq C\delta^\alpha$$

Since E is bounded, this stability theorem tells us that $\left| E \triangle B_\rho^+ \right|$ becomes smaller and smaller as $\varepsilon \to 0$. To obtain uniform convergence we have to use the regularity properties of E. By Lemma 2.1 or 2.2, if there was one point of ∂E far from ∂B_ρ^+, then there would be a fixed amount of volume of $E \triangle B_\rho^+$ around it, arriving to a contradiction. We state the theorem:

Theorem 2.3. *Given any $\eta > 0$, for ε small enough*

$$B_{(1-\eta)\rho}^+ \subset E \subset B_{(1+\eta)\rho}^+$$

2.5 Hysteresis

Although when we consider absolute minimizers of J_ε there are no surprises in the homogenization limit, in reality this behavior is almost never observed. When a drop is formed, its shape does not necessarily achieve an absolute minimum of the energy, but it stabilizes in any local minimum of J_ε. That is why to fully understand the possible shapes of drops lying on a rough surface, we must study the limits as $\varepsilon \to 0$ of all the critical points of J_ε.

Let us see a simplified equation in 1 dimension. Let u be the solution of the following free boundary problem:

$$u \geq 0 \qquad \text{in } [0, 1]$$
$$u(0) = 0$$
$$u(1) = 1$$
$$u''(x) = 0 \qquad \text{if } u(x) > 0$$
$$\frac{du}{dx^+} = \beta\left(\frac{x}{\varepsilon}\right) \quad \text{for } x \in \partial\{u > 0\}$$

$$\tan \gamma = \beta$$

This problem comes from minimizing the functional

$$J(u) = \int_0^1 |u'|^2 + \beta\left(\frac{x}{\varepsilon}\right)^2 \chi_{u>0} \, dx$$

If β is constant, it is clear that there is only one solution, because only one line from $(1, 1)$ hits the x axis with an angle $\gamma = \arctan \beta$. However, if β oscillates, there must be several solutions that correspond to several critical points of J_ε. There will be a solution hitting the x axis at the point x_0 as long as $\frac{1}{1-x_0} = \beta\left(\frac{x_0}{\varepsilon}\right)$. For a small ε this may happen at many points, as we can see in Figure 7.

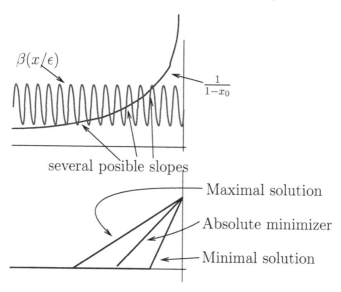

$\beta(x/\epsilon)$

$\dfrac{1}{1-x_0}$

several posible slopes

Maximal solution

Absolute minimizer

Minimal solution

Fig. 7. Different solutions for a nonconstant β

Moreover, the set of possible slopes for the solutions gets more and more dense in the interval $[\min \beta, \max \beta]$ as ε gets small. As $\varepsilon \to 0$, we can get a sequence of solutions converging to a segment with any slope in that interval.

This example shows that the situation is not so simple. When we go back to our problem of the drop in more than one dimension, the expected possible slopes as $\varepsilon \to 0$ must be in the interval $[\arccos \max \beta, \arccos \min \beta]$. Exactly what they are depends on the particular geometry of the problem. If for example β depends only on one variable, let us say x_1, then when the free boundary aligns with the direction of x_1 we would expect to obtain a whole range of admissible slopes as in the $1D$ case. Let us sketch a proof in this case that there is a sequence of critical points of the functional that do not converge to a sphere cap as $\varepsilon \to 0$. We will construct a couple of barriers, and then find solutions that stay below them.

Suppose that β depends on only one variable and it is not constant. As we have shown in the previous section, the absolute minimizers converge to a sphere cap B_ρ^+ as $\varepsilon \to 0$. Let $S(x_1)$ be a function that touches B_ρ^+ at one end point $x_1 = -R$, but has a steeper slope at that point. Let us choose this slope $S'(-R) = \tan \alpha$ such that $\cos \alpha < \max \beta$, we can do this from the extra room that we have since β is not constant. Now let us continue $S(x_1)$ from that point first with a constant curvature larger than the curvature of B_ρ^+, and then continued as linear. Since S starts off with a steeper slope than B_ρ^+, we can make S so that $S > B_\rho^+$ for $x_1 > -R$. Now we translate S a tiny bit in the direction of x_1 to obtain S_1 so that $S_1 \leq B_\rho^+$ only in the set where S_1 has a positive curvature that is larger than the one of B_ρ^+. We construct a similar function S_2 in the other side of B_ρ^+. See Figure 8.

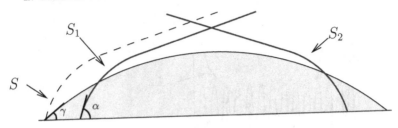

Fig. 8. Barrier functions

We will see that we can find a sequence of solutions for $\varepsilon \to 0$ that remains under S_1 and S_2. For suitable choices of ε, $\cos \alpha < \beta(-R)$ and also $\cos \alpha < \beta(R)$. For such ε, we minimize the energy J_ε constrained to remain below S_1 and S_2. In other words, we minimize J_ε from all the sets E subsets of

$$D = \{(x_1, x') : -R \leq x_1 \leq R \wedge z \leq S_1(x_1) \wedge z \leq S_2(x_1)\}$$

If E is the constrained minimizer, it will be a critical point (unconstrained) of J_ε as long as it does not touch the graphs of S_1 or S_2. Since only a tiny bit of B_ρ^+ is outside of D, $J_\varepsilon(E)$ will not differ from $J(B_\rho^+)$ much when ε is small. We can then apply the stability result of section 2.4 to deduce that ∂E remains in a neighborhood of ∂B_ρ^+. The curvature of ∂E will be constant where it is a free surface, and no larger than that value where it touches the boundary of D. Since ∂E is close to ∂B_ρ^+ everywhere, the curvature of the free part of ∂E cannot be very different from the curvature of ∂B_ρ^+. Therefore E cannot touch S_1 or S_2 in the part where these barriers are curved. The part where these barriers are straight is too far away from B_ρ^+, so E cannot reach that part either. It is only left to check the boundary $x_1 = \pm R$ and $z = 0$. But the contact angle of S_1 is smaller than $\arccos \beta(x_1)$ at those points, and then E cannot reach those points either. Thus, E must be a free minimizer. Since we can do this for ε arbitrarily small, when $\varepsilon \to 0$ we obtain limits of the homogenization problem that cannot be the sphere cap B_ρ^+ because they are trapped in a narrower strip $\{-R + \delta \leq x_1 \leq R + \delta\}$.

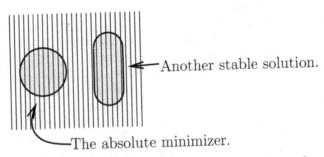

Another stable solution.

The absolute minimizer.

Fig. 9. Different drops can be formed on irregular surfaces

Other geometries may produce different variations. It is hard to predict what can be expected.

We may ask at this point what is then the shape that we will observe in a real physical drop. The answer is that it depends on how it was formed. If the equilibrium was reached after an expansion, then we can expect to see the largest possible contact angle. If on the other hand, the equilibrium was obtained after for example evaporation, then we can expect to see the least possible contact angle.

An interesting case is the drop lying on an inclined surface. If we consider gravity, there is no absolute minimizer for the energy, because we can slide down the drop all the way down and make the energy tend to $-\infty$. However, we see drops sitting on inclined surfaces all the time. The reason is that they stabilize in critical points for the energy. On the side that points down, we can see a larger contact angle than the one in the other side. This effect would not be possible in a ideal perfectly smooth surface.

2.6 References

The equations of capillarity can be found in [12]. The case of constant β is studied in [G].

The proof of Theorem 2.2, as well as the existence of a minimizer for each ε and a comprehensive development of the topic can be found in [5].

Lemma 2.2 is not as in [5]. There a different approach is taken that also leads to Corollaries 2.1 and 2.2. This modification was suggested by several people.

The phenomena of Hysteresis, and in particular the case of the drop on an inclined surface is discussed in [6]. Previous references for hysteresis are [17], [16] and [15].

Related methods are used for the problem of flame propagation in periodic media [3], [4].

3 The Construction of Plane Like Solutions to Periodic Minimal Surface Equations

The second homogenization problem that we would like to discuss is related to minimal surfaces in a periodic medium.

In two dimensions, minimal surfaces are just geodesics. Suppose we are given a differential of length $a(x, \nu)$ in $\mathbb{R}^2$, and given two points x, y we want to find the curve joining them with the minimum possible length. In other words, we want to minimize

$$d(x, y) = \inf L(\gamma) = \int_\gamma a(z, \sigma)\, ds$$

among all curves γ joining x to y.

Here s is the usual differential of length and σ the unit tangent vector. We consider a function $a(x, \sigma)$ that is strictly positive ($0 < \lambda \leq a(x, \sigma) \leq \Lambda$) and, to avoid the formation of Young measures (that is: oscillatory zig-zags) when trying to construct geodesics, it must satisfy

$$|v| a\left(x, \frac{v}{|v|}\right) \text{ is a strictly convex cone.}$$

We assume that a is periodic in unit cubes. By that we mean that a is invariant under integer translations, i.e. $a(x + h, \sigma) = a(x, \sigma)$ for any vector h with integer coordinates. Let us also assume that a is smooth although this property is not needed. Due to the periodicity, at large distances $d(x, y)$ becomes almost translation invariant, since for any vector z there is a vector $\tilde{z}$ with integer coordinates such that $|z - \tilde{z}| \leq \frac{\sqrt{n}}{2}$ and

$$|d(x + z, y + z) - d(x, y)| = |d(x + z, y + z) - d(x + \tilde{z}, y + \tilde{z})|$$
$$\leq \sqrt{n}\,\Lambda$$

Another way of saying the same thing is to look at the geodesics from very far away, that is to rescale the medium by a very small ε,

$$a_\varepsilon(x, \sigma) = a\left(\frac{x}{\varepsilon}, \sigma\right).$$

The distance becomes almost translation invariant

$$|d_\varepsilon(x, y) - d_\varepsilon(x + z, y + z)| \leq \varepsilon\sqrt{n}\,\Lambda$$

and as ε goes to zero we obtain an effective norm $\|x\| = \lim_{\varepsilon \to 0} d_\varepsilon(x, 0)$.

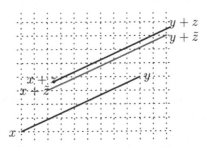

Fig. 10. The distance is almost translation invariant

The question we are interested to study is the following: given any line

$$L = \{\lambda\sigma, \lambda \in R\}$$

Can we construct a global geodesic S that stays at a finite distance from L? That is S remains trapped in a strip, around L whose width depends only on λ, Λ.

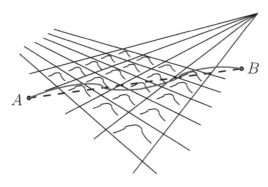

Fig. 11. Line like geodesic

The answer is *yes* in 2D (Morse) and *no* in 3D (Hevlund). An inspection of Hevlund counterexample shows that, unlike classical homogenization, where diffusion processes tend to *average the medium*, geodesics try to *beat the medium* by choosing specific paths, and leaving *bad areas* untouched.

In the 80's Moser suggested that in $\mathbb{R}^n$, unlike geodesics, minimal hypersurfaces should be forced to *average* the medium, and given any plane π, it should be possible to construct *plane like* minimal surfaces for the periodic medium.

More precisely given a *differential of area* form, we would like to consider surfaces S that locally minimize

Fig. 12. Hevlund Counterexample: It costs *one* to travel inside narrow pipes, a large K outside. Then, the best strategy is to jump only once from pipe to pipe, i.e., the effective norm is $\|x\| = |x| + |y| + |z|$

$$A^*(S) = \int_S a(x, \nu) \, dA$$

where dA is the usual differential of area, ν the normal vector to A, and a, as before satisfies,

i. $0 < \lambda \le a(x, \nu) \le \Lambda$
ii. $|v|a(x, v/|v|)$ is a strictly convex cone.
iii. a is periodic in x.

These conditions for a, translate in the following properties of A^*.

i. $\lambda Area(S) \le A^*(S) \le \Lambda Area(S)$.
ii. $A^*(S) = A^*(\tau_z S)$, for any translation τ_z with integer coordinates.

By a local minimizer of A^*, we mean a surface S such that if another surface S_1 coincides with S everywhere but in a bounded set B, then $A^*(S \cap B) \le A^*(S_1 \cap B)$.

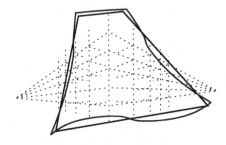

Fig. 13. Plane-like minimal surface in a periodic medium (for instance a medium with a periodic Riemman metric)

The main theorem is the following:

Theorem 3.1. *There exists a universal constant $M(\lambda, \Lambda, n)$ such that: for any unit vector ν_0 there exists an A^* local area minimizer S contained in the strip $\pi_M = \{x : |\langle x, \nu_0 \rangle| < M\}$.*

A first attempt to construct such local area minimizer is to look at surfaces that are obtained by adding a periodic perturbation to the plane $\pi = \{x : \langle x, \nu_0 \rangle = 0\}$. This will be possible if π has a rational slope, or equivalently that π can be generated by a set of $n - 1$ vectors $e_1, \dots, e_{n-1}$ with integer coordinates. The advantage of this case is that a translation in the direction of each e_j fixes π as well as the metric, so we can expect that we can find a local A^* minimizer that is also fixed by the same set of translations. If we can prove Theorem 3.1 in this context and the constant M does not depend on the vectors $e_1, \dots, e_{n-1}$ but only on λ, Λ and dimension, then the general case (irrational slope) follows by a limiting process.

We will work in the framework of boundaries of sets of locally finite perimeter.

A set of locally finite perimeter Ω is a set such that for any ball B, $B \cap \Omega$ has a finite perimeter (as in the first part of these notes, see [13]) For such sets, differential of area of $\partial\Omega$, and unit normal vectors are well defined, under our hypothesis A^* makes sense and is lower semicontinuous under convergence in measure for sets.

Main Steps of the Proof

We will consider the family of sets $\mathbb{D}$ such that $\Omega \in \mathbb{D}$ if Ω is a set of locally finite perimeter, $\tau_{e_j}(\Omega) = \Omega$ for every j (where $\tau_{e_j}(\Omega) := \Omega + e_j$), and

$$\pi_M^- = \{x : \langle x, \nu_0 \rangle \geq -M\} \subset \Omega \subset \pi_M^+ = \{x : \langle x, \nu_0 \rangle \leq M\}$$

And within $\mathbb{D}$, we will consider those sets Ω_0 that are local A^*-minimizers among sets $\Omega \in \mathbb{D}$. Since we are in the context of periodic perturbations of a plane, a local A^*-minimizer is simply a minimizer of A^* of the portion of $\partial\Omega$ inside the fundamental cube given by all the points of the form $\lambda_1 e_1 + \cdots + \lambda_{n-1}e_{n-1} + \lambda_n \nu_0$ where $\lambda_j \in [0,1]$ for $j = 1, \ldots, n-1$ and $\lambda_n \in [-M, M]$.

Of course, such an Ω_0 is not a *free* local minimizer since whenever $\partial\Omega_0$ touches the boundary of π^- or π^+ we are not free to perturb it outwards.

Our objective is to show that if M is large enough $S_0 = \partial\Omega_0$ does not see this restriction. In other words, Ω_0 would be a local A^*-minimizer not only among the sets in $\mathbb{D}$ but also among all sets of locally finite perimeter.

The main ingredients are:

a) A positive density property
b) An area estimate for $\partial\Omega$

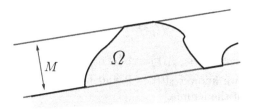

Fig. 14. Restricted Minimizer

c) Minimizers are *ordered*

Lemma 3.1 (Positive density). *There are two universal constants $c_0, C_1 > 0$ such that a minimizer $\partial\Omega_0$ of A^* satisfies*

$$c_0 r^n \leq \frac{|\Omega_0 \cap B_r(x_0)|}{|B_r|} \leq C_1 r^n$$

for any $x_0 \in \partial\Omega_0$.

Proof. This lemma is actually the same as Lemma 2.1 in a slightly different context. The only difference is that instead of (3), we must use now that $\partial \Omega_0$ is a minimal surface. We include the proof here for completeness.

We define

$$U_1(r) = |B_r(x_0, z_0) \setminus \Omega_0| \qquad S_1(r) = Area(\partial B_r(x_0, z_0) \setminus \Omega_0)$$

$$U_2(r) = |B_r(x_0, z_0) \cap \Omega_0| \qquad S_2(r) = Area(\partial B_r(x_0, z_0) \cap \Omega_0)$$

$$A(r) = Area(B_r \cap \partial \Omega_0)$$

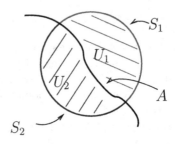

Since $\partial \Omega_0$ is a minimal surface,

$$A(r) \leq \frac{1}{\lambda} A^*(B_r \cap \partial \Omega_0) \leq \frac{1}{\lambda} A^*(\partial B_r(x_0, z_0) \setminus \Omega_0) \leq \frac{\Lambda}{\lambda} S_1(r)$$

Similarly $A(r) \leq \frac{\Lambda}{\lambda} S_2(r)$.

We also know by the isoperimetrical inequality that $U_1^{\frac{n}{n+1}} \leq C(A + S_1)$. If we combine this with the above inequality we obtain

$$U_1^{\frac{n}{n+1}} \leq C S_1$$

But now we observe that $S_1(r) = U_1'(r)$, so we obtain the ODE: $U_1'(r) \geq c U_1^{\frac{n}{n+1}}$. Moreover, we know $U_1(0) = 0$ and $U_1(r) > 0$ for any $r > 0$. This implies the result of the lemma.

In the same way, we obtain the result for U_2.

Lemma 3.2. *There are two universal constants $c_0, C_1 > 0$ such that a minimizer $\partial \Omega_0$ of A^* satisfies*

$$c_0 R^{n-1} \leq \mathcal{H}^{n-1}(\partial \Omega_0 \cap B_R) \leq C_1 R^{n-1}$$

for large values of R.

Proof. Notice that the set $\Omega_1 = \{x : \langle x, \nu \rangle < 0\}$ is an admissible set in $\mathbb{D}$. Then $A^*(\partial \Omega_0 \cap$ fundamental cube$) \leq A^*(\partial \Omega_1 \cap$ fundamental cube$)$. Besides, $Area(\partial \Omega_1 \cap$ fundamental cube$) \leq Area(\partial \Omega_0 \cap$ fundamental cube$)$. Thus, $Area(\partial \Omega_0 \cap B_R)$ and $Area(\partial \Omega_1 \cap B_R)$ are comparable when R is large.

This would be the same as the result of the lemma if it was true that the area of the boundary of a set of finite perimeter coincides with its $n - 1$ Hausdorff measure. Unfortunately, that is not always true. In general we can say that the $n - 1$ Hausdorff measure is only greater or equal to the area. But in this case we can compare them thanks to Lemma 3.1. If we take a finite overlapping covering with balls of radius r centered at $\partial \Omega_0 \cap B_R$, by Lemma 3.1 plus the isoperimetric inequality, the surface of $\partial \Omega_0$ inside each ball cannot be less than $c_0 r^{n-1}$. Then, there cannot be more than $C R^{n-1} / r^{n-1}$ such balls, and the Hausdorff estimate follows.

Lemma 3.3. *Minimizers are* ordered, *that is if Ω_0 and Ω_1 are minimizers, then so are $\Omega_0 \cup \Omega_1$ and $\Omega_0 \cap \Omega_1$.*

Proof. $\partial \Omega_0 \cup \partial \Omega_1 = \partial(\Omega_0 \cup \Omega_1) \cup \partial(\Omega_0 \cap \Omega_1)$ and thus if we add the areas (A^*) inside the fundamental cube of $\partial \Omega_0$ and $\partial \Omega_1$, it is the same as adding the corresponding ones for $\partial(\Omega_0 \cap \Omega_1)$ and $\partial(\Omega_0 \cup \Omega_1)$. But since Ω_0 and Ω_1 are A^* area minimizers, necessarily all those areas are the same and then both $\Omega_0 \cup \Omega_1$ and $\Omega_0 \cap \Omega_1$ must be minimizers too.

Using Lemma 3.3, we can construct the smallest minimizer $\overline{\Omega}$ in $\mathbb{D}$ by taking the intersection of all minimizers in $\mathbb{D}$. We point out the similarity with Perron's method.

$\overline{\Omega}$ recuperates an important property, the Birkhoff property: If τ_z is an integer translation with $\langle z, \nu_0 \rangle \leq 0$ (resp. ≥ 0) then

$$\tau_z(\overline{\Omega}) \subset \overline{\Omega} \quad (\text{resp.} \; \supset \overline{\Omega})$$

Indeed $\tau_z(\overline{\Omega}) \cap \overline{\Omega}$ and $\tau_z(\overline{\Omega}) \cup \overline{\Omega}$ are minimizers respectively for $\tau_z(\pi_M)$ and π_M, while $\overline{\Omega}$ and $\tau_z(\overline{\Omega})$ are the actual smallest minimizers.

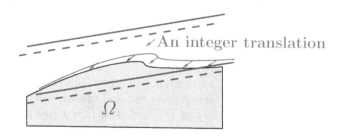

Fig. 15. Birkhoff Property. Integer translations send $\tau_z(\Omega)$ inside Ω or Ω inside $\tau_z(\Omega)$ depending on whether $\langle z, \nu_0 \rangle \leq 0$ or $\langle z, \nu_0 \rangle \geq 0$

Lemma 3.2 tells us that for large balls $B_R(0)$, the number N of disjoint unit cubes intersecting $\partial \Omega_0$ must be of order $N \sim C_1 R^{n-1}$ independently of M. Since the strip $\pi_M \cap B_R$ has roughly $M R^{n-1}$ cubes, many cubes in $\pi_M \cap B_R$ must be contained in Ω_0 or $\mathcal{C}\Omega_0$.

Combining the above properties we see the following:

i) There are many *clean cubes* that do not intersect $\partial\overline{\Omega}$, and thus they are contained in either $\overline{\Omega}$ or its complement. Moreover, there are many such cubes that are not too close to the boundary of π_M.

ii) Any integer translation $\tau_z(Q)$ of a cube $Q \subset \overline{\Omega}$ with $\langle z, \nu_0 \rangle \le 0$ is contained in $\overline{\Omega}$. Conversely for a cube $Q \subset \mathcal{C}\overline{\Omega}$, if $\langle z, \nu_0 \rangle \ge 0$ then $\tau_z(Q) \subset \mathcal{C}\overline{\Omega}$.

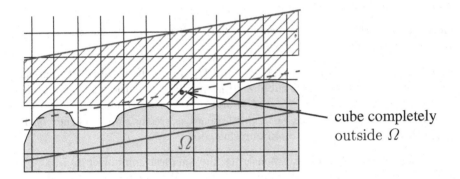

cube completely outside Ω

Fig. 16. If one cube is outside of $\overline{\Omega}$, then any cube whose center is above the dotted line is outside of $\overline{\Omega}$

From i), we can find a *clean* cube Q that is not too close to the boundary of π_M. If this cube Q is contained in $\overline{\Omega}$ and M is large, then the union of all the translations $\tau_z(Q)$ for z with integer coordinates and $\langle z, \nu_0 \rangle \le 0$ covers a strip around the bottom of π_M (see Figure 16 *upside down*). But then we have a thick *clean* strip, which means that we could translate $\overline{\Omega}$ a unit distance down and still have a local minimizer, which would contradict the fact that $\overline{\Omega}$ is the minimum of them.

Therefore, we must be able to find a *clean* cube contained in $\mathcal{C}\overline{\Omega}$. Arguing as above, this implies that there is a complete *clean* strip around the top of π_M (like in Figure 16). Thus, we are free to perturb upwards. Moreover, we can lift the whole set $\overline{\Omega}$ by an integer amount and obtain another minimizer that does not touch the boundary of π_M, and then $\overline{\Omega}$ is a *free* minimizer.

In this way we prove the theorem when π has a rational slope. Since M depends only on λ, Λ and dimension, we approximate a general π by planes with rational slopes and prove the theorem by taking the limit of the respective minimizers (or a subsequence of them).

3.1 References

The content of this part is based on the joint paper with Rafael de la Llave [2].

The problem had been proposed by Moser in another C.I.M.E. course [M1] (See also [M2], [18]). The interest of constructing line like geodesics was related to foliating the torus with them or at least laminate it.

4 Existence of Homogenization Limits for Fully Nonlinear Equations

Let us start the third part of these notes with a review on the definitions of fully nonlinear elliptic equations.

A second order fully nonlinear equation is given by an expression of the form

$$F(D^2u, Du, u, x) = 0 \tag{5}$$

for a general nonlinear function $F : \mathbb{R}^{n \times n} \times \mathbb{R}^n \times \mathbb{R} \times \mathbb{R}^n \to \mathbb{R}$. For simplicity, we will consider equations that do not depend on Du or u. So they have the form

$$F(D^2u, x) = 0 \tag{6}$$

The equation (6) is said to be *elliptic* when $F(M + N, x) \geq F(M, x)$ every time N is a positive definite matrix. Moreover, (6) is said to be *uniformly elliptic* when we have $\lambda |N| \leq F(M+N, x) - F(M, x) \leq \Lambda |N|$ for two positive constants $0 < \lambda \leq \Lambda$ and where $|N|$ denotes the norm of the matrix N. The simplest example of a uniformly elliptic equation is the laplacian, for which $F(M, x) = \operatorname{tr} M$.

Existence, uniqueness and regularity theory for uniformly elliptic equations is a well developed subjet. It is studied in the framework of viscosity solutions that is a concept that was first introduced by Crandall and Lions for Hamilton Jacobi equations. We will consider only uniformly elliptic equations throughout this section.

A continuous function u is said to be a *viscosity* subsolution of (6) in an open set Ω, and we write $F(D^2u, x) \geq 0$, when each time a second order polynomial P touches u from above at a point $x_0 \in \Omega$ (i.e. $P(x_0) = u(x_0)$ and $P(x) > u(x)$ for x in a neighborhood of x_0), then $F(D^2P(x_0), x_0) \geq 0$. Respectively, u is a supersolution $(F(D^2u, x) \leq 0)$ if every time P touches u from below at x_0 then $F(D^2P(x_0), x_0) \leq 0$. For the general theory of viscosity solutions see [8] or [1].

In the same way as for subharmonic and superharmonic functions, sub- and supersolutions of uniformly elliptic equations satisfy the comparison principle: if u and v are respectively a sub- and supersolution of an equation like (6) and $u \leq v$ on the boundary of a bounded domain Ω, then also $u \leq v$ in the interior of Ω.

Suppose now that we have a family of uniformly elliptic equations (with the same λ and Λ) that do not depend on x (are translation invariant): $F_j(D^2u) = 0$ for $j = 1, \ldots, k$. Let us suppose that at every point in space we choose one of these equations with some probability. To fix ideas, let us divide $\mathbb{R}^n$ into unit cubes with integer corners and in each cube we pick one of these equations at random with some given probability. The equation that we obtain for the whole space will change on each cube, it will not look homogeneous, it will not be translation invariant, and it will strongly depend on the random choice at every cube. However if we look at the equation from far away, somehow the differences from point to point should average out and we should obtain a translation invariant equation.

In each black square we have $F_1(D^2u) = 0$
In each white square we have $F_2(D^2u) = 0$

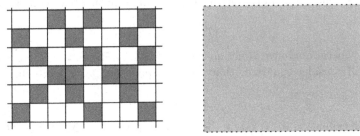

From close, we see black and white squares   From far, we just see gray

Fig. 17. A chessboard like configuration

Let $(\mathbb{S}, \mu)$ be the probability space of all the possible configuration. For each $\omega \in \mathbb{S}$ we have an x-dependent equation

$$F(D^2u, x, w) = 0$$

What we would expect is that if we consider solutions u_ω^ε of the equation (with same given boundary values)

$$F(D^2u_\omega^\varepsilon, \frac{x}{\varepsilon}, w) = 0 \tag{7}$$

with probability 1, they would converge to solutions u_0 of a translation invariant (constant coefficients) equation

$$F(D^2u_0) = 0$$

thus, in this limiting process that corresponds to looking at the medium from far away, the differences from point to point should dissapear. An moreover, it should lead to the same uniform equation for almost all ω.

Our purpose is to prove the existence of this limiting equation.

The appropriate setting for the idea of mixed media that from far away looks homogeneous is ergodic theory. Out assumptions are:

1. For each w in the probability space $\mathbb{S}, \mu$ we have a uniformly elliptic equation

$$F(D^2 u, x, w) = 0$$

defined in all $\mathbb{R}^n$.

2. Translating the equation in any direction z with integer coordinates is the same as shifting the configuration w, i.e.

$$F(M, x - z, w) = F(M, x, \tau_z(w))$$

and we ask this transformation $w \mapsto \tau_z(w)$ to preserve probability.

3. *Ergodicity assumption*: For any set $S \subset \mathbb{S}$ of positive measure, the union of all the integer translations of S covers almost all $\mathbb{S}$

$$\mu \left(\bigcup_{z \in \mathbb{Z}^n} \tau_z(S) \right) = 1$$

Under these conditions, we obtain the following theorem:

Theorem 4.1. *There exists an homogenization limit equation*

$$\tilde{F}(D^2 u_0) = 0$$

to which solutions of the problem (7) converge almost surely.

4.1 Main Ideas of the Proof

When we have a translation invariant equation $F(D^2 u) = 0$, if u is a solution of such equation, that means that for each point x, the matrix $D^2 u(x)$ lies on the zero level set $\{M \in \mathbb{R}^{n \times n} : F(M) = 0\}$. We can describe the equation completely if we are able to classify all quadratic polynomials P as solutions, subsolutions or supersolutions, because that would tell us for what matrices M, $F(M)$ is equal, greater or less than zero.

Let us choose a polynomial P_0 in a large cube Q_R and let us compare $P_0 + t|x|^2$ with the solution of

$$F(D^2 u, x, w) = 0 \qquad\qquad \text{in } Q_R$$
$$u = P_0 + t|x|^2 \qquad\qquad \text{in } \partial Q_R$$

If t is very large, $P_0 + t|x|^2$ will be a subsolution of the equation and thus $P_0 + t|x|^2 \leq u$ in Q_R. Equally, if λ is very negative then $P_0 + t|x|^2 \geq u$ in Q_R. For some intermediate values of t, $P_0 + t|x|^2$ and u cross each other, so for these values it is not so clear at this point if $P_0 + t|x|^2$ is going to be a sub or supersolution of the homogenization limit equation.

Let us forget about the term $t\,|x|^2$ for a moment. Given a quadratic polynomial $P = \sum_{ij} M_{ij} x_i x_j$, we want to solve the equation

$$F(D^2 u^\varepsilon, \frac{x}{\varepsilon}, w) = 0 \quad \text{in } Q_1$$

$$u^\varepsilon = P \quad \text{on } \partial Q_1 \tag{8}$$

for a unit cube Q_1. Subsolutions of our homogenized equations are those polynomials for which u^ε tends to lie above P as $\varepsilon \to 0$. Similarly, supersolutions are those for which u^ε tends to be below P. If the polynomial P is borderline between these two behaviors, then it would be a solution of the homogenization limit equation.

It is important to notice that we can either think of the problem at scale ε in a unit cube (with u^ε) or we can keep unit scale and consider a large cube. To look at the equation (8) for $\varepsilon \to 0$ is equivalent to keep the same scale and consider larger cubes. Indeed, if we consider $u(x) = \frac{1}{\varepsilon^2} u^\varepsilon(\varepsilon x)$, then for $R = \varepsilon^{-1}$, we have

$$F(D^2 u, x, w) = 0 \quad \text{in } Q_R$$

$$u = P \quad \text{in } \partial Q_R \tag{9}$$

For a cube Q_R of side R. It is convenient to choose R to be integer, in order to fit an integer number of whole unit cubes in Q_R. Now instead of taking $\varepsilon \to 0$, we can take $R \to +\infty$. We will be switching between these two points of view constantly.

Let v be the solution of the corresponding obstacle problem. The function v is the least supersolution of the equation (9) such that $v \geq P$:

$$F(D^2 v, x, w) \leq 0 \quad \text{in } Q_R$$

$$v = P \quad \text{in } \partial Q_R$$

$$v \geq P \quad \text{in } Q_R \tag{10}$$

$$F(D^2 v, x, w) = 0 \quad \text{in the set } \{v > P\}$$

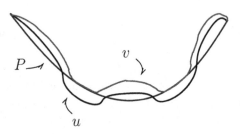

Fig. 18. The polynomial P, the free solution u and the least supersolution above the polynomial v

We also call $v^\varepsilon = \varepsilon^2 v(x/\varepsilon)$, the solution of the obstacle problem at scale ε. Let ρ be the measure of the contact set $\{v = P\}$ in Q_R:

$$\rho(Q_R) = |\{v = P\}|$$

The value of ρ controls the difference between u and v. A small value of ρ means that v touches P at very few points, and thus it is *almost* a free solution. The idea is that if ρ remains small compared to $|Q_R|$ as $R \to +\infty$, then P would be a subsolution of the homogenized equation. A large value of ρ means that v touches P in many points. If $\frac{\rho}{|Q_R|} \to 1$ as $R \to \infty$, that would mean that P is a supersolution. Moreover, we will show that every time $\frac{\rho}{|Q_R|}$ converges to a positive value, then $u^\varepsilon \to P$.

The first thing we must prove is that $\frac{\rho}{|Q_R|}$ indeed converges to some value as $R \to +\infty$ (or $\varepsilon \to 0$). Notice that $\frac{\rho}{|Q_R|}$ is the measure of the contact set at scale ε: $|\{v^\varepsilon = P\}|$.

In this problem, what plays the role of the Birkhoff property is a subadditivity condition for ρ, as the following lemma says.

Lemma 4.1. *If a cube Q is the disjoint union of a sequence of cubes Q_j, then*

$$\rho(Q) \le \sum_j \rho(Q_j)$$

Proof. Let v be the solution of the obstacle problem in the cube Q that coincides with P on ∂Q. Let v_j be the corresponding ones for the cubes Q_j. Since $v \ge P$ in Q, $v \ge v_j$ on ∂Q_j. Then by comparison principle $v \ge v_j$ in Q_j. Therefore the contact set $\{x \in Q : v(x) = P(x)\}$ is contained in the union of the contact sets $\{x \in Q_j : v_j(x) = P(x)\}$, and the lemma follows.

This subadditivity condition plus the ergodicity condition and

$$\rho(Q_R(x - z), \omega) = \rho(Q_R(x), \tau_z(\omega))$$

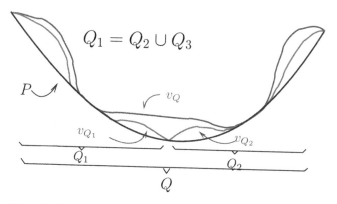

Fig. 19. Pay attention to the contact sets: ρ is subadditive

are the conditions for a subadditive ergodic theorem (which can be found in [9]) that says that as R go to infinity $\frac{\rho(Q_R(x_0))}{|Q_R(x_0)|}$ converges to a constant h_0 with probability 1. We will characterize polynomials P as sub- or supersolutions according to whether $h_0 = 0$ or $h_0 > 0$.

Lemma 4.2. *If $h_0 = 0$, then*

$$\liminf_{\varepsilon \to 0} u^\varepsilon \geq P$$

Proof. Using the Alexandrov-Backelman-Pucci inequality (See for example [1]), we can obtain a precise estimate of $v^\varepsilon - u^\varepsilon$ depending on ρ:

$$\sup_{Q_R} v^\varepsilon - u^\varepsilon \leq CR\rho^{1/n}$$

where C is a universal constant.

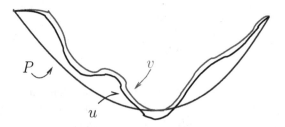

Fig. 20. If the contact set if small, then u and v are close

If $h_0 = 0$, as ε goes to zero we have:

$$u^\varepsilon(x) \geq v^\varepsilon(x) - C\varepsilon\rho^{1/n}$$

$$\geq v^\varepsilon(x) - C\left(\frac{\rho}{|Q_R|}\right)^{1/n}$$

$$\geq P - o(1)$$

Then, as $\varepsilon \to 0$, u^ε tends to be above P, and we finish the proof of the lemma.

The last lemma suggests that P is a subsolution of the homogenization limit equation if $h_0 = 0$. Now we will consider the case $h_0 > 0$. In order to show that in that case u_ε tends to be below P, we have to use that v^ε separates from P by a universal quadratic speed depending only on the ellipticity of the equation.

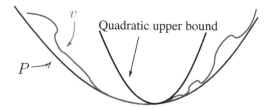

Fig. 21. Quadratic separation

The quadratic separation from the contact set is a general characteristic of the obstacle problem. What it means is that if $v^\varepsilon(x_0) = P(x_0)$, then

$$v^\varepsilon(x) - P(x) \le C\,|x - x_0|^2$$

for a constant C depending only on λ, Λ and dimension.

The quadratic separation in this problem plays the role of the positive density in the previous ones.

Lemma 4.3. *If $h_0 > 0$, then*

$$\limsup_{\varepsilon \to 0} u^\varepsilon \le P$$

Proof. We will show that the contact set $\{v^\varepsilon = P\}$ spreads all over the unit cube. Then, using the quadratic separation we show that $v^\varepsilon \to P$ as $\varepsilon \to 0$.

We want to show that if we split the unit cube in m smaller cubes of equal size, for any value of m, then for ε small enough there is a piece of the contact set in each small cube. We know that the measure of the contact set $|\{x \in Q_1 : v^\varepsilon(x) = P(x)\}|$ converges to $h_0 > 0$. The unit cube Q_1 is split into m smaller cubes. Let Q be any of these cubes, we have $v^\varepsilon \ge P$ on ∂Q, so v^ε is a supersolution of the corresponding obstacle problem in Q and $\frac{|\{x \in Q : v^\varepsilon = P(x)\}|}{|Q|}$ cannot converge to any value larger than h_0 as $\varepsilon \to 0$. If in some cube the contact set is empty $\{x \in Q : v^\varepsilon(x) = P(x)\} = \varnothing$, then, since the whole contact set covers a proportion h_0 of the measure of the unit cube, there must be one of the smaller cubes where the contact set covers more than h_0 times the measure of this cube (at least for a sequence $\varepsilon_k \to 0$). And that is a contradiction, which means that the contact set $\{v^\varepsilon = P\}$ must spread all over.

But if $\{v^\varepsilon = P\}$ spreads all over the unit cube, then v^ε converges to P uniformly due to the universal quadratic separation. Since $v^\varepsilon \ge u^\varepsilon$, $\limsup_{\varepsilon \to 0} u^\varepsilon \le P$.

So, now we have a way to classify every polynomial as subsolution to the homogenization limit equation $(\tilde{F}(D^2P) \ge 0)$ if $h_0 = 0$ or supersolution $(\tilde{F}(D^2P) \le 0)$ if $h_0 > 0$. There is still a little bit of ambiguity because a polynomial could be both things at a time (if it is precisely a solution). That is easily solved by considering $P_0 + t\,|x|^2$ for small values of t. We say that P_0

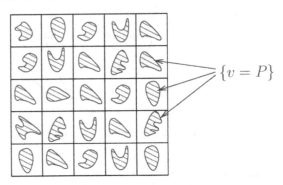

Fig. 22. Each small cube must contain about the same amount of contact set when $\varepsilon \ll 1$

is a sub or supersolution if we can check it for $P_0 + t\,|x|^2$ for arbitrarily small values of t.

In this way we are able to completely characterize the zero level set of $\tilde{F}$.

Moreover, if we want to construct the complete function $\tilde{F}$, then we have to identify all its level sets, not only the zero level set. To do that we just consider the problem:

$$F(D^2 u^\varepsilon, \frac{x}{\varepsilon}, w) - t = 0$$

to describe the level set $\tilde{F}(M) = t$. And we recover $\tilde{F}$ completely.

Now, based on our construction of $\tilde{F}$, it is easy to show that for any boundary data, problem (7) will converge with probability 1 to a function u_0 that satisfies comparison with polynomials in the right way to be a viscosity solution of $\tilde{F}(D^2 u_0) = 0$. We finish with the theorem:

Theorem 4.2. *Let u^ε be the solutions to*

$$F(D^2 u^\varepsilon, \frac{x}{\varepsilon}, w) = 0 \quad in \ \Omega$$

$$u^\varepsilon = g \quad in \ \partial\Omega$$

(11)

for a domain Ω and a continuous function g on $\partial\Omega$. Then as $\varepsilon \to 0$, almost surely u^ε converge uniformly to a function u that solves

$$\tilde{F}(D^2 u) = 0 \quad in \ \Omega$$

$$u = g \quad in \ \partial\Omega$$

(12)

Proof. Due to the uniform ellipticity of F, the functions u^ε are uniformly continuous, and therefore by Arzela-Ascoli there is a subsequence u^{ε_k} that converges uniformly to a continuous function u.

Let us suppose that a quadratic polynomial P touches u from above at a point x_0. Then we can lower P a little bit by subtracting a small constant δ_1

such that $P(x_0) < u(x_0)$ and $P(x) > u(x)$ for x in the boundary of a small cube $Q_{\delta_2}(x_0)$ centered at x_0.

Since $u^{\varepsilon k}$ converge to u uniformly, the same property holds for them. Namely, for large enough k

$$P(x_0) \leq u^{\varepsilon k}(x_0) - \delta_1$$

$$P(x) > u^{\varepsilon k}(x) \quad \text{for } x \in \partial Q_{\delta_2}(x_0)$$

Let w_k be the solutions to

$$F(D^2 w_k, \frac{x}{\varepsilon}, w) = 0 \quad \text{in } Q_{\delta_2}(x_0)$$

$$w_k = P \quad \text{in } \partial Q_{\delta_2}(x_0)$$

(13)

By comparison principle, $w_k \leq u^{\varepsilon k}$, then $w_k(x_0) \leq P(x_0) - \delta_1$ for large k. So, we can apply Lemma 4.3 to obtain that the value of h_0 corresponding to P cannot be positive. Then $\tilde{F}(D^2 P) \geq 0$.

In a similar way, we can show that if a quadratic polynomial touches u from below then it must be a supersolution of $\tilde{F}$.

Therefore u must be a viscosity solution of (12). Since (12) has a unique solution, all the convergent subsequences of u^ε must converge to the same limit. Thus the whole sequence u^ε converges uniformly to u.

4.2 References

This part in homogenization is based on the joint work with Panagiotis Souganidis and Lihe Wang [7], where actually a more complete theorem is proved. The fact that equations that depend on ∇u are considered in that paper adds some extra complications.

Some of the ideas have their roots in the work of Dal Maso and Modica ([9] and [10]) for the variational case.

Periodic homogenization for second order elliptic equations was considered in [11].

Acknowledgements

These notes were partially written during the period Luis Silvestre was employed by the Clay Mathematics Institute as a Liftoff Fellow.

Luis Caffarelli was partially supported by NSF grants.

The authors would also like to thank the organizers of the C.I.M.E. course *Calculus of variations and nonlinear partial differential equations* for their hospitality.

References

1. Luis A. Caffarelli and Xavier Cabré. *Fully nonlinear elliptic equations*, volume 43 of *American Mathematical Society Colloquium Publications*. American Mathematical Society, Providence, RI, 1995.
2. Luis A. Caffarelli and Rafael de la Llave. Planelike minimizers in periodic media. *Comm. Pure Appl. Math.*, 54(12):1403–1441, 2001.
3. Luis A. Caffarelli, Ki-Ahm Lee, and Antoine Mellet. Singular limit and homogenization for flame propagation in periodic excitable media. *Arch. Ration. Mech. Anal.*, 172(2):153–190, 2004.
4. Luis A. Caffarelli, Ki-Ahm Lee, and Antoine Mellet. Homogenization and flame propagation in periodic excitable media: the asymptotic speed of propagation. *Comm. in Pure and Appl. Math.*, To appear.
5. Luis A. Caffarelli and Antoine Mellet. Capillary drops on an inhomogeneous surface. *Preprint*, 2005.
6. Luis A. Caffarelli and Antoine Mellet. Capillary drops on an inhomogeneous surface: Contact angle hysteresis. *Preprint*, 2005.
7. Luis A. Caffarelli, Panagiotis E. Souganidis, and Lihe Wang. Homogenization of fully nonlinear, uniformly elliptic and parabolic partial differential equations in stationary ergodic media. *Comm. Pure Appl. Math.*, 58(3):319–361, 2005.
8. Michael G. Crandall, Hitoshi Ishii, and Pierre-Louis Lions. User's guide to viscosity solutions of second order partial differential equations. *Bull. Amer. Math. Soc. (N.S.)*, 27(1):1–67, 1992.
9. Gianni Dal Maso and Luciano Modica. Nonlinear stochastic homogenization. *Ann. Mat. Pura Appl. (4)*, 144:347–389, 1986.
10. Gianni Dal Maso and Luciano Modica. Nonlinear stochastic homogenization and ergodic theory. *J. Reine Angew. Math.*, 368:28–42, 1986.
11. Lawrence C. Evans. Periodic homogenisation of certain fully nonlinear partial differential equations. *Proc. Roy. Soc. Edinburgh Sect. A*, 120(3-4):245–265, 1992.
12. Robert Finn. *Equilibrium capillary surfaces*, volume 284 of *Grundlehren der Mathematischen Wissenschaften [Fundamental Principles of Mathematical Sciences]*. Springer-Verlag, New York, 1986.
13. Enrico Giusti. *Minimal surfaces and functions of bounded variation*, volume 80 of *Monographs in Mathematics*. Birkhäuser Verlag, Basel, 1984.
14. Eduardo H. A. Gonzalez. Sul problema della goccia appoggiata. *Rend. Sem. Mat. Univ. Padova*, 55:289–302, 1976.
15. C Huh and S.G Mason. Effects of surface roughness on wetting (theoretical). *J. Colloid Interface Sci*, 60:11–38.
16. J.F. Joanny and P.G. de Gennes. A model for contact angle hysteresis. *J. Chem. Phys.*, 81, 1984.
17. L. Leger and J.F. Joanny. Liquid spreading. *Rep. Prog. Phys.*, pages 431–486, 1992.
18. Jürgen Moser. Minimal solutions of variational problems on a torus. *Ann. Inst. H. Poincaré Anal. Non Linéaire*, 3(3):229–272, 1986.
19. Jürgen Moser. A stability theorem for minimal foliations on a torus. *Ergodic Theory Dynam. Systems*, 8*(Charles Conley Memorial Issue):251–281, 1988.
20. Jürgen Moser. Minimal foliations on a torus. In *Topics in calculus of variations (Montecatini Terme, 1987)*, volume 1365 of *Lecture Notes in Math.*, pages 62–99. Springer, Berlin, 1989.

A Visit with the ∞-Laplace Equation

Michael G. Crandall*

Department of Mathematics, University of California, Santa Barbara,
Santa Barbara, CA 93106, USA
crandall@math.ucsb.edu

Introduction

In these notes we present an outline of the theory of the archetypal L^∞ variational problem in the calculus of variations. Namely, given an open $U \subset \mathbb{R}^n$ and $b \in C(\partial U)$, find $u \in C(\overline{U})$ which agrees with the boundary function b on ∂U and minimizes

$$\mathcal{F}_\infty(u, U) := \||Du|\|_{L^\infty(U)} \tag{0.1}$$

among all such functions. Here $|Du|$ is the Euclidean length of the gradient Du of u. We will also be interested in the "Lipschitz constant" functional as well. If K is any subset of $\mathbb{R}^n$ and $u \colon K \to \mathbb{R}$, its least Lipschitz constant is denoted by

$$\mathrm{Lip}(u, K) := \inf \left\{ L \in \mathbb{R} : |u(x) - u(y)| \le L|x - y| \; \forall \, x, y \in K \right\}. \tag{0.2}$$

Of course, $\inf \emptyset = +\infty$. Likewise, if any definition such as (0.1) is applied to a function for which it does not clearly make sense, then we take the right-hand side to be $+\infty$. One has $\mathcal{F}_\infty(u, U) = \mathrm{Lip}(u, U)$ if U is convex, but equality does not hold in general.

Example 2.1 and Exercise 2.1 below show that there may be many minimizers of $\mathcal{F}_\infty(\cdot, U)$ or $\mathrm{Lip}(\cdot, U)$ in the class of functions agreeing with a given boundary function b on ∂U. While this sort of nonuniqueness can only take place if the functional involved is not strictly convex, it is more significant here that the functionals are "not local." Let us explain what we mean in contrast with the Dirichlet functional

$$\mathcal{F}_2(u, U) := \frac{1}{2} \int_U |Du|^2 \, dx. \tag{0.3}$$

This functional has the property that if u minimizes it in the class of functions satisfying $u|_{\partial U} = b$ and $V \subset U$, then u minimizes $\mathcal{F}_2(\cdot, V)$ among functions

---

* Supported in part by NSF Grant DMS-0400674.

which agree with u on ∂V, properly interpreted. This is what we mean by "local" here. Exercise 2.1 establishes that both Lip and $\mathcal{F}_\infty$ are not local, as you can also do with a moment's thought.

This lack of locality can be rectified by a notion which directly builds in locality. Given a general nonnegative functional $\mathcal{F}(u, V)$ which makes sense for each open subset V of the domain U of u, one says that $u : U \to \mathbb{R}$ is *absolutely minimizing*[2] for $\mathcal{F}$ on U provided that

$$\mathcal{F}(u, V) \le \mathcal{F}(v, V) \text{ for every } v : \overline{V} \to \mathbb{R} \text{ such that } u = v \text{ on } \partial U.$$

whenever $\overline{V}$ is compactly contained in the domain of u. Of course, we need to supplement this idea with some more precision, depending on $\mathcal{F}$, but we won't worry about that here. Clearly, if u is absolutely minimizing for $\mathcal{F}$ on U, then it is absolutely minimizing for $\mathcal{F}$ on open subsets of U. The absolutely minimizing notion decouples the considerations from the boundary condition; it defines a class of functions without regard to behavior at the boundary of U. We might then consider the problem: find $u : \overline{U} \to \mathbb{R}$ such that

$$u \text{ is absolutely minimizing for } \mathcal{F} \text{ on } U \text{ and } u = b \text{ on } \partial U. \tag{0.4}$$

It is not quite clear that a solution of this problem minimizes $\mathcal{F}$ among functions which agree with b on ∂U. Indeed, it is not true for $\mathcal{F} = \mathcal{F}_\infty, \text{Lip}, \mathcal{F}_2$ if U is unbounded (Exercise 2.1). The notion also does not require the existence of a function u satisfying the boundary condition for which $\mathcal{F}(u, U) < \infty$.

The theory of absolutely minimizing functions and the problem (0.4) for the functional Lip is quicker and slicker than that for $\mathcal{F}_\infty$, and we present this first, ignoring $\mathcal{F}_\infty$ for a while. However, it is shown in Section 6 that a function which is absolutely minimizing for Lip is also absolutely minimizing for $\mathcal{F}_\infty$ and conversely. It turns out that the absolutely minimizing functions for Lip and $\mathcal{F}_\infty$ are precisely the viscosity solutions of the famous partial differential equation

$$\Delta_\infty u = \sum_{i,j=1}^n u_{x_i} u_{x_j} u_{x_i x_j} = 0. \tag{0.5}$$

The operator Δ_∞ is called the "∞-Laplacian" and solutions of (0.5) are said to be ∞-harmonic. The reason for this nomenclature is given in Section 8. The notion of a "viscosity solution" is given in Definition 2.4 and again in Section 8, together with more information about them than is in the main text.

An all-important class if ∞-harmonic functions (equivalently, absolutely minimizing functions for $\mathcal{F} = \mathcal{F}_\infty$ or $\mathcal{F} = \text{Lip}$) is the class of *cone functions*:

$$C(x) = a|x - z|, \tag{0.6}$$

where $a \in \mathbb{R}$ and $z \in \mathbb{R}^n$. It turns out that the ∞-harmonic functions are precisely those that have a comparison property with respect to cone

---

[2] The author prefers the more descriptive term "locally minimizing," but c'est la vie.

functions. All the basic theory of ∞-harmonic functions can be derived by *comparison with cones*. This is explained and exploited in Sections 2–7.

The table of contents gives an impression of how these notes are organized, and we will not belabor that here, except for some comments. Likewise, the reader will have noticed a lack of references in this introduction. We will give only two, including the research/expository article [8]. These notes are closely related to [8]. However, [8] treats the situation in which $|\cdot|$ is a general norm on $\mathbb{R}^n$ rather than the Euclidean norm. There is perhaps a cost in elegance in this generality, and we welcome the opportunity to write the story of the Euclidean case by itself. In particular, at the time of this writing, it is not quite settled whether or not the absolutely minimizing property is equivalent to a partial differential equation in the case of a general norm. It has, however, been shown that this is true for, for example, the l_∞ and l_1 norms on $\mathbb{R}^n$, which was unknown at the writing of [8]. There is a rather complete set of references in [8], along with comments, up to the time it was written. We rely partly on [8] in this regard.

In Section 8 we give an informal outline of the nearly 40 year long saga of the theory of the ∞-Laplace equation. This section is intended to be readable immediately after this introduction; it corresponds to a talk the author gave at a conference in honor of G. Aronsson, the initiator of the theory, in 2004. Selected references are given in Section 8. In addition, in Section 9, we attempt to give a feeling for the many generalizations and the scope of recent activity in this active area, going well beyond the basic case we study here in some detail, including sufficient references and pointers to provide the interested reader entree into whatever part of the evolving landscape suits their interests.

What is new in the current article, relative to [8], besides many details of the organization and presentation and Sections 8, 9? Quite a number of things, of which we point out the direct derivation of the ∞-Laplace equation in the viscosity sense from comparison with cones in Section 2.2 (a refinement of an original argument in [8]), the gradient flow curves in Section 6, the outline of a new uniqueness proof in Section 5, a new result in Section 7, a variety of items near the end of Section 4 and a little proof we'd like to share in Section 7.2. In addition, the current text is supplemented by exercises. Most of them fall in the range from straightforward to very easy and of some of them are solved in the main text of [8]. Whether or not the reader attempts them, they should be read.

Regarding our exposition, we build in some redundancy to ease the flow of reading. In particular, Section 8 is largely accessible now, after this introduction.

In the lectures to which these notes correspond, the author spent considerable time on a recent result of O. Savin [49]. Savin proved that ∞-harmonic functions are C^1 if $n = 2$. This was a big event. Whether or not this is true if $n > 2$, and the author would bet that it is, remains the most prominent open problem in the area. The author had hoped to include an exposition of Savin's results here, but did not in the end do so, due to lack of time. In any case, any such exposition would have been very close to the original article. If it is proved that ∞-harmonic functions are C^1 in general, this article will need substantial revision; however, most of the theory is about ∞-subharmonic functions, which are not C^1, so it is unlikely the material herein will be rendered obsolete. In any case, we'd best get on with it!

1 Notation

In these notes, U, V, W are always open subsets of $\mathbb{R}^n$. The closure of U is $\overline{U}$ and its boundary is ∂U. The statement $V \ll U$ means that $\overline{V}$ is a compact subset of U.

If $x = (x_1, \ldots, x_n), y = (y_1, \ldots, y_n) \in \mathbb{R}^n$, then

$$|x| := \left(\sum_{j=1}^{n} x_j^2 \right)^{1/2} \quad \text{and} \quad \langle x, y \rangle := x_1 y_1 + \cdots + x_n y_n.$$

The notation $A := B$ means that A is defined to be B. If $K \subset \mathbb{R}^n$ and $x \in \mathbb{R}^n$, then

$$\text{dist}\,(x, K) := \inf_{y \in K} |x - y|.$$

If $x^j \in \mathbb{R}^n$, $j = 1, 2, \ldots$, then

$$x^j \to y \quad \text{means} \quad y \in \mathbb{R}^n \text{ and } \lim_{j \to \infty} x^j = y.$$

The space of continuous real valued functions on a topological space K is denoted by $C(K)$. The notation "$u \in C^k$" indicates that u is a real-valued function on a subset of $\mathbb{R}^n$ and it is k-times continuously differentiable. $L^\infty(U)$ is the standard space of essentially bounded Lebesgue measurable function with the usual norm,

$$\|v\|_{L^\infty(U)} = \inf \left\{ M \in \mathbb{R} : |v(x)| \leq M \text{ a.e in } U \right\}.$$

We also use pointwise differentiability and twice differentiability. For example, if $w : U \to \mathbb{R}$ and $y \in U$, then w is twice differentiable at y if there exist $p \in \mathbb{R}^n$ and a real symmetric $n \times n$ matrix such that

$$w(x) = w(y) + \langle p, x - y \rangle + \frac{1}{2} \langle X(x - y), x - y \rangle + o(|x - y|^2). \tag{1.1}$$

In this event, we write $Du(y) = p$ and $D^2u(y) = X$.

Exercise 1.1. Show that (1.1) can hold for at most one pair p, X. If (1.1) holds and y is a local maximum point for w, then $p = 0$ and $X \leq 0$ (in the usual ordering of symmetric matrices: $X \leq 0$ iff $\langle X\zeta, \zeta \rangle \leq 0$ for $\zeta \in \mathbb{R}^n$).

Balls are denoted as follows:

$$B_r(x) := \{y \in \mathbb{R}^n : |y - x| < r\}, \ \overline{B}_r(x) := \{y \in \mathbb{R}^n : |y - x| \le r\},$$

If $w, z \subset \mathbb{R}^n$, then

$$[w, z] := \{w + t(z - w) : 0 \le t \le 1\}$$

is the line segment from w to z. Similarly, $(w, z) := \{w + t(z - w) : 0 < t < 1\}$ and so on.

2 The Lipschitz Extension/Variational Problem

Let $b \in C(\partial U)$. We begin by considering the problem: find u such that

$$\begin{cases} u \in C(\overline{U}), u = b \text{ on } \partial U \text{ and} \\ \mathrm{Lip}(u, \overline{U}) = \min \left\{ \mathrm{Lip}(v, \overline{U}) : v \in C(\overline{U}), v = b \text{ on } \partial U \right\} \end{cases} \quad (2.1)$$

The notation "b" for the boundary data above is intended as a mnemonic. It is clear that if $u \in C(\overline{U})$, then $\mathrm{Lip}(u, \partial U) \le \mathrm{Lip}(u, \overline{U}) = \mathrm{Lip}(u, U)$. Thus, if $\mathrm{Lip}(b, \partial U) = \infty$, any continuous extension of b into U is a solution of (2.1). Moreover, if $\mathrm{Lip}(b, \partial U) < \infty$ and $u \in C(\overline{U})$ agrees with b on ∂U, then $\mathrm{Lip}(u, U) = \mathrm{Lip}(b, \partial U)$ guarantees that u solves (2.1).

Assuming that $\mathrm{Lip}(b, \partial U) < \infty$, it is easy to see that (2.1) has a maximal and a minimal solution which in fact satisfy $\mathrm{Lip}(u, U) = \mathrm{Lip}(b, \partial U)$. Indeed, if $\mathrm{Lip}(u, U) = \mathrm{Lip}(b, \partial U)$, $z, y \in \partial U$, $x \in U$ and $L = \mathrm{Lip}(b, \partial U)$, then u must satisfy

$$b(z) - L|x - z| = u(z) - L|x - z| \le u(x) \text{ and}$$
$$u(x) \le u(y) + L|x - y| = b(y) + L|x - y|.$$

This implies

$$\sup_{z \in \partial U} (b(z) - L|x - z|) \le u(x) \le \inf_{y \in \partial U} (b(y) + L|x - y|). \quad (2.2)$$

Denote the left-hand side of (2.2) by $\mathcal{MW}_*(b)(x)$ and the right hand side of (2.2) by $\mathcal{MW}^*(b)(x)$. The notation is in honor of McShane and Whitney, see Section 8. Since infs and sups over functions with a given Lipschitz constant possess the same Lipschitz constant, $\mathrm{Lip}(\mathcal{MW}_*(b), \mathbb{R}^n), \mathrm{Lip}(\mathcal{MW}^*(b), \mathbb{R}^n) \le L = \mathrm{Lip}(b, \partial U)$. It is also obvious that $\mathcal{MW}^*(b) = \mathcal{MW}_*(b) = b$ on ∂U. Thus $\mathcal{MW}^*(b)$ $(\mathcal{MW}_*(b))$ provides a maximal (respectively, minimal) solution of (2.1). Since $\mathcal{MW}^*(b), \mathcal{MW}_*(b)$ have the same Lipschitz constant as b, they are regarding as solving the "Lipschitz extension problem," in that they extend b to $\mathbb{R}^n$ while preserving the Lipschitz constant. "Extension" has different overtones than does "variational."

There is no reason for these extremal solutions to coincide, and it is rare that they do. The example below shows this, no matter how nice U might be.

Example 2.1. Let $U = \{x \in \mathbb{R}^2 : |x| < 1\}$ be the unit disc in $\mathbb{R}^2$ and $b \in C(\partial U)$. Arrange that $-1 \le b \le 1$; this can be done with arbitrarily large $L = \text{Lip}(b, \partial U)$. Then, via (2.2), $\mathcal{MW}_*(b)(0) < \mathcal{MW}^*(b)(0)$ provided that there is a $\delta > 0$ for which

$$b(z) - L|z| + \delta = b(z) - L + \delta$$
$$< b(y) + L|y| = b(y) + L \text{ for } |y| = |z| = 1.$$

Since $-1 \le b \le 1$, this is satisfied if $L > 1$, for then

$$\delta - 2L \le -1 - 1 = -2 \le b(y) - b(z)$$

with $\delta = (L - 1)$.

A primary difference between the functional Lip and more standard integral functionals is that Lip is not "local," as explained in the introduction. The next exercise shows this.

Exercise 2.1. Let $n = 1$, $U = (-1, 0) \cup (0, 1)$ and $b(-1) = b(0) = 0$, $b(1) = 1$. Find $\mathcal{MW}^*(b)$ and $\mathcal{MW}_*(b)$. Conclude that the functional Lip is not local. Modify this example to show that even if U is bounded, then it is not necessarily true that $\mathcal{MW}^*(b) \le \max_{\partial U} b$ in U, nor does $b \le \tilde{b}$ necessarily imply that $\mathcal{MW}^*(b) \le \mathcal{MW}^*(\tilde{b})$ in U (for example).

We recall the localized version of the idea that u "minimizes the functional Lip."

Definition 2.1. *Let* $u : U \to \mathbb{R}$. *Then is an absolute minimizer for the functional* Lip *on* U *provided that* $u \in C(U)$ *and*

$$\text{Lip}(u, V) \le \text{Lip}(v, V) \text{ whenever } V \ll U, v \in C(\overline{V}) \text{ and } u = v \text{ on } \partial V.$$
$$(2.3)$$

We will use various ways to refer to this notion, saying, equivalently, that "u is an absolute minimizer for Lip on U," or "u is absolutely minimizing for Lip on U," or "u is absolutely minimizing Lipschitz" or merely writing

$$u \in \text{AML}(U).$$

The notion is evidently local in the sense that if $u \in \text{AML}(U)$ and $V \subset U$, then $u \in \text{AML}(V)$.

Exercise 2.2. With the notation of Exercise 2.1, determine all the continuous functions on $\mathbb{R}$ and all the continuous functions on $[-1, 1]$ which agree with b on ∂U and which are absolute minimizers for Lip on $\mathbb{R} \setminus \partial U$ and U, respectively. More generally, show that if $\mathcal{I}$ is an interval and $u \in C(\mathcal{I})$ is an absolute minimizer for Lip on $\mathcal{I}$, then u is linear.

Note that the notion of an absolute minimizer does not involve boundary conditions; it is a property of functions defined on open sets. We recast the problem (2.1) and in terms of this notion. This results in: find u with the properties

$$u \in C(\overline{U}) \cap \mathrm{AML}(U) \text{ such that } u = b \text{ on } \partial U. \tag{2.4}$$

It is not clear that (2.4) has any solutions, whether or not $\mathrm{Lip}(b, \partial U) = \infty$. Nor is it clear that a solution is unique if it exists. Your solution of Exercise 2.2 shows that solutions are not unique in general. We will see in Section 5 that solutions exist very generally and are unique if U is bounded.

Theorem 2.1. *The following are equivalent conditions on a function* $u \in C(U)$.

(a) $u \in \mathrm{AML}(U)$.
(b) If $w = u$ *or* $w = -u$, *then for every* $a \in \mathbb{R}$, $V \ll U$ *and* $z \notin V$

$$w(x) - a|x - z| \leq \max_{y \in \partial V} (w(y) - a|y - z|) \text{ for } x \in V. \tag{2.5}$$

(c) If $w = u$ *or* $w = -u$ *and* $\varphi \in C^2(U)$ *and* $w - \varphi$ *has a local maximum at* $\hat{x} \in U$, *then*

$$\Delta_\infty \varphi(\hat{x}) := \sum_{i,j=1}^{n} \varphi_{x_i}(\hat{x}) \varphi_{x_j}(\hat{x}) \varphi_{x_i x_j}(\hat{x}) \geq 0. \tag{2.6}$$

Definition 2.2. *If* $a \in \mathbb{R}$ *and* $z \in \mathbb{R}^n$, *we call the function* $C(x) = a|x - z|$ *a* cone function. *The slope of* C *is* a *and its vertex is* z. *The half-line* $\{z + t(x - z), t \geq 0\}$ *is the* ray *of* C *through* x.

Definition 2.3. *A function* $w \in C(U)$ *with the property (2.5) is said to enjoy* comparison with cones from above *in* U. *If* $-w$ *enjoys comparison with cones from above, equivalently*

$$w(x) - C(x) \geq \max_{y \in \partial V} (w(y) - C(y)) \text{ for } x \in V$$

for $V \ll U$ *and cone functions* C *whose vertices are not in* V, w *is said to enjoy* comparison with cones from below. *If* w *enjoys comparison with cones from above and from below, then it enjoys* comparison with cones. *Thus, condition (b) of Theorem 2.1 is that "u enjoys comparion with cones."*

Definition 2.4. *The differential operator given by*

$$\Delta_\infty \varphi := \sum_{i,j=1}^{n} \varphi_{x_i} \varphi_{x_j} \varphi_{x_i x_j} = \langle D^2 \varphi D\varphi, D\varphi \rangle \tag{2.7}$$

on smooth functions φ *is called the "*∞*-Laplacian. Here* $D\varphi = (\varphi_{x_1}, \ldots, \varphi_{x_n})$ *is the gradient of* φ *and* $D^2\varphi = (\varphi_{x_i x_j})$ *is the Hessian matrix of second* *derivatives of φ. A function $w \in C(U)$ such that (2.6) holds for every φ in $C^2(U)$ and local maximum $\hat{x}$ of $w - \varphi$ is said to be a viscosity subsolution of $\Delta_\infty w = 0$; equivalently, it is a viscosity solution of $\Delta_\infty w \geq 0$ or w is ∞-subharmonic. If $-w$ is a ∞-subharmonic, equivalently, at any local minimum $\hat{x} \in U$ of $w - \varphi$ where $\varphi \in C^2$, one has*

$$\Delta_\infty \varphi(\hat{x}) \leq 0,$$

then w is ∞-superharmonic. If w is both ∞-subharmonic and ∞- superharmonic, then it is ∞-harmonic and we write $\Delta_\infty w = 0$.

Important Notice: Hereafter the modifier "viscosity" will be often be dropped, *as was already done in defining, for example, "∞ - subharmonic".*

The viscosity notions are the right ones here, and are taken as primary. One does not, in general, compute the expression "$\Delta_\infty w$" and evaluate it, as in $\Delta_\infty w(x)$, to determine whether or not u is ∞-harmonic. Instead, one checks the conditions of the definition above, or some equivalent, as in Theorem 2.1. However, the expression $\Delta_\infty w(x)$ does have a pointwise meaning if w is twice differentiable at x, that is,

$$w(z) = w(x) + \langle p, z - x \rangle + \frac{1}{2} \langle X(z - x), z - x \rangle + o(|z - x|^2) \qquad (2.8)$$

for some $p \in \mathbb{R}^n$ and real symmetric $n \times n$ matrix X. Then

$$\Delta_\infty w(x) = \langle D^2 w(x) Dw(x), Dw(x) \rangle = \langle Xp, p \rangle.$$

Noting, for example, that if (2.8) holds, $\varphi \in C^2$, and $w - \varphi$ has a maximum at x, then $p = D\varphi(x)$ and $X \leq D^2\varphi(x)$ (see Exercise 1.1), we find that

$$\Delta_\infty w(x) = \langle Xp, p \rangle = \langle X D\varphi(x), D\varphi(x) \rangle \leq \langle D^2\varphi(x) D\varphi(x), D\varphi(x) \rangle.$$

It follows that if $w \in C^2$, then $\Delta_\infty w \geq 0$ in the pointwise sense implies $\Delta_\infty w \geq 0$ in the viscosity sense. Similarly, if (2.8) holds, then for $\varepsilon > 0$

$$z \mapsto w(z) - \left(w(x) + \langle p, z - x \rangle + \frac{1}{2} \langle (X + \varepsilon I)(z - x), z - x \rangle \right)$$

has a maximum at $z = x$, so if w is a viscosity solution of $\Delta_\infty w \geq 0$, we must have $\langle (X + \varepsilon I)p, p \rangle \geq 0$. Letting $\varepsilon \downarrow 0$, we find $\Delta_\infty w(x) \geq 0$. Thus the viscosity notions are entirely consistent with the pointwise notion at points of twice differentiability.

Remark 2.1. In Exercises 2.6, 2.7 below you will show that the function defined on $\mathbb{R}^2$ by $u(x, y) = x^{4/3} - y^{4/3}$ is ∞-harmonic on $\mathbb{R}^2$. As it is not twice differentiable on the coordinate axes, this cannot be checked via pointwise computation of $\Delta_\infty u$. The viscosity notions, which we are taking as primary here, give a precise meaning to the claim that $\Delta_\infty u = 0$.

We break out the proof of Theorem 2.1 in several simple parts.

2.1 Absolutely Minimizing Lipschitz iff Comparison With Cones

We begin with a useful triviality. If $g\colon [a,b] \subset \mathbb{R} \to \mathbb{R}$, then

$$\operatorname{Lip}(g,[a,b])| = \frac{|g(b) - g(a)|}{|b - a|} \implies$$
$$g(a + t(b - a)) = g(a) + t(g(b) - g(a)) \text{ for } 0 \le t \le 1. \tag{2.9}$$

This has the following obvious consequence: If $u\colon [w,z] \subset \mathbb{R}^n \to \mathbb{R}$, then

$$\frac{|u(z) - u(w)|}{|z - w|} = \operatorname{Lip}(u,[w,z]) \implies$$
$$u(w + t(z - w)) = u(w) + t(u(z) - u(w)) \text{ for } 0 \le t \le 1. \tag{2.10}$$

Assume now that $u \in C(U)$ enjoys comparison with cones (Definition (2.3)). Note that the comparison with cones from above is equivalent to the condition (2.11) (a) below. Likewise, comparison with cones from below may be restated as (2.11) (b). If $a, c \in \mathbb{R}$ and $z \notin V$, then

(a) $u(x) \le c + a|x - z|$ for all $x \in V$ if it holds for $x \in \partial V$,

(b) $c + a|x - z| \le u(x)$ for all $x \in V$ if it holds for $x \in \partial V$. $\tag{2.11}$

We show that, for any $x \in V$,

$$\operatorname{Lip}(u, \partial(V \setminus \{x\})) = \operatorname{Lip}(u, \partial V \cup \{x\}) = \operatorname{Lip}(u, \partial V). \tag{2.12}$$

To see this we need only check that if $y \in \partial V$, then

$$u(y) - \operatorname{Lip}(u, \partial V)|x - y| \le u(x) \le u(y) + \operatorname{Lip}(u, \partial V)|x - y|. \tag{2.13}$$

As each of the above inequalities holds for $x \in \partial V$ and u enjoys comparison with cones, the inequalities indeed hold if $x \in V$. Let $x, y \in V$. Using (2.12) twice,

$$\operatorname{Lip}(u, \partial V) = \operatorname{Lip}(u, \partial(V \setminus \{x\})) = \operatorname{Lip}(u, \partial(V \setminus \{x,y\})).$$

Since $x, y \in \partial(V \setminus \{x,y\})$, we have $|u(x) - u(y)| \le \operatorname{Lip}(u, \partial V)|x - y|$, and hence $u \in \operatorname{AML}(U)$.

Let $C(x) = a|x - z|$ be a cone function. Notice that $\operatorname{Lip}(C, [w,y]) = |a|$ whenever w, y are distinct points on the same ray of C. Thus $\operatorname{Lip}(C, V) = |a|$ for any nonempty open set V and $\operatorname{Lip}(C, \partial V) = |a|$ if V is bounded, nonempty and does not contain the vertex z of C.

Suppose now that $u \in \operatorname{AML}(U)$. Assume that $V \ll U$, $z \notin V$ and set

$$W = \{x \in V : u(x) - a|x - z| > \max_{w \in \partial V} (u(w) - a|w - z|)\}. \tag{2.14}$$

We want to show that W is empty. If it is not empty, then it is open and

$$u(x) = a|x - z| + \max_{w \in \partial V}(u(w) - a|w - z|) =: C(x) \text{ for } x \in \partial W. \tag{2.15}$$

Therefore $u = C$ on ∂W and $\mathrm{Lip}(u, W) = \mathrm{Lip}(C, \partial W) = |a|$ since u is absolutely minimizing. Now if $x_0 \in W$, the ray of C through x^0, $t \mapsto z + t(x_0 - z)$, $t \geq 0$, contains a segment in W containing x_0 which meets ∂W at its endpoints. Since $t \mapsto C(z + t(x_0 - z)) = at|x_0 - z|$ is linear on this segment, with slope $a|x_0 - z|$, while $t \mapsto u(z + t(x_0 - z))$ also has $|a||x_0 - z|$ as a Lipschitz constant and the same values at the endpoints of the segment; therefore it is the same function ((2.10)). Thus

$$u(z + t(x_0 - z)) = C(z + t(x_0 - z))$$

on the segment, which contains x^0, whence $u(x_0) = C(x_0)$, a contradiction to $x_0 \in W$. Thus W is empty.

Remark 2.2. In showing that comparison with cones implies AML, we only used comparison with cones with nonnegative slopes from above, and comparison with cones with nonpositive slopes from below.

Exercise 2.3. Let $u \in C(U)$ be absolutely minimizing for $\mathcal{F}_\infty$, that is, whenever $V \ll U$, $v \in C(\overline{V})$ and $u = v$ on ∂V, then $\mathcal{F}_\infty(u, V) \leq \mathcal{F}_\infty(v, V)$. Show that u enjoys comparison with cones. Hint: the proof above needs only minor tweaking.

2.2 Comparison With Cones Implies ∞-Harmonic

You may prefer other proofs, but the one below is direct; it does not use contradiction. First we use comparison with cones from above, which implies that, using the form (2.11) (a) of this condition,

$$u(x) \leq u(y) + \max_{\{w:|w-y|=r\}} \left(\frac{u(w) - u(y)}{r} \right) |x - y| \qquad (2.16)$$

for $x \in B_r(y) \ll U$. The inequality (2.16) holds as asserted because it trivially holds for $x \in \partial(B_r(y) \setminus \{y\})$.

Rewrite (2.16) as

$$u(x) - u(y) \leq \max_{\{w:|w-y|=r\}} (u(w) - u(x)) \frac{|x - y|}{r - |x - y|} \qquad (2.17)$$

for $x \in B_r(y) \ll U$. If u is twice differentiable at x, namely, if there is a $p \in \mathbb{R}^n$ and a symmetric $n \times n$ matrix X such that

$$u(z) = u(x) + \langle p, z - x \rangle + \frac{1}{2} \langle X(z - x), z - x \rangle + \mathrm{o}\left(|z - x|^2\right),$$
$$\text{so that } p := Du(x),\ X := D^2 u(x), \qquad (2.18)$$

we will show that

$$\Delta_\infty u(x) = \langle D^2 u(x) Du(x), Du(x) \rangle = \langle Xp, p \rangle \geq 0. \qquad (2.19)$$

That is, comparison with cones from above implies $\Delta_\infty u \geq 0$ at points of twice differentiability.

We are going to plug (2.18) into (2.17) with two choices of z. First, on the left of (2.17), we choose $z = y = x - \lambda p$ where p is from (2.18), and expand $u(x) - u(y)$ according to (2.18). Next, let $w_{r,\lambda}$ be a value of w for which the maximum on the right of (2.17) is attained and expand $u(w_{r,\lambda}) - u(x)$ according to (2.18). This yields, after dividing by $\lambda > 0$,

$$|p|^2 + \lambda \frac{1}{2} \langle Xp, p \rangle + o(\lambda)$$

$$\leq \left(\langle p, w_{r,\lambda} - x \rangle + \frac{1}{2} \langle X(w_{r,\lambda} - x), w_{r,\lambda} - x \rangle + o((r+\lambda)^2) \right) \frac{|p|}{r - \lambda |p|}$$

$$(2.20)$$

Sending $\lambda \downarrow 0$ yields

$$|p|^2 \leq \left(\left\langle p, \frac{w_r - x}{r} \right\rangle + \frac{1}{2} \left\langle X \left(\frac{w_r - x}{r} \right), w_r - x \right\rangle \right) |p| + |p| o(r)$$

$$\leq |p|^2 + \frac{1}{2} \left\langle X \left(\frac{w_r - x}{r} \right), w_r - x \right\rangle |p| + |p| o(r),$$

$$(2.21)$$

where w_r is a any limit point of the $w_{r,\lambda}$ as $\lambda \downarrow 0$ and therefore $w_r \in \partial B_r(x)$ - so $(w_r - x)/r$ is a unit vector. Since the second term inside the parentheses on the right of the first inequality above has size r and $(w_r - x)/r$ is a unit vector, it follows from the first inequality that $(w_r - x)/r \to p/|p|$ as $r \downarrow 0$. (We are assuming that $p \neq 0$, as we may.) Then the inequality of the extremes in (2.21), after dividing by r and letting $r \downarrow 0$, yields $0 \leq \langle Xp, p \rangle$, as desired.

The above proof contains a bit more information than $0 \leq \langle Xp, p \rangle$ if $p = Du(x) = 0$. In this case, choosing y so that $|x - y| = r/2$, we have

$$u(x) - u(y) = O(r^2)$$

and then (2.17) yields

$$O(r^2) \leq \frac{1}{2} \left\langle X \left(\frac{w_{r,y} - x}{r} \right), w_{r,y} - x \right\rangle + o(r)$$

where $(w_{r,y} - y)/r$ is a unit vector. Dividing by r, sending $r \downarrow 0$ and using compactness, any limit point of $(w_{r,y} - x)/r$ as $r \downarrow 0$ is a unit vector q for which $0 \leq \langle Xq, q \rangle$. In particular, if $Du(x) = 0$, then

$$D^2 u(x) \quad \text{has a nonnegative eigenvalue.} \tag{2.22}$$

This set up is a bit more awkward than is necessary for the results obtained so far. It is set up this way to make the next remark easy. If x is a local maximum point of $u - \varphi$ for some smooth φ, then

$$\varphi(x) - \varphi(y) \leq u(x) - u(y) \quad \text{and} \quad u(w) - u(x) \leq \varphi(w) - \varphi(x)$$

for y, w near x. That is, we may replace u by φ in (2.17). By what was just shown, it follows that $\Delta_\infty \varphi(x) \geq 0$. That is, by definition, u is a viscosity solution of $\Delta_\infty u \geq 0$ if it satisfies comparison with cones from above. We record this again: if u enjoys comparison with cones from above, then

$$\varphi \in C^2, u - \varphi \text{ has a local max at } x \implies \Delta_\infty \varphi(x) \geq 0. \qquad (2.23)$$

In addition, if $D\varphi(x) = 0$, then

$$D^2 \varphi(x) \quad \text{has a nonnegative eigenvalue.} \qquad (2.24)$$

Similarly, if u enjoys comparison with cones from below, then

$$\varphi \in C^2, u - \varphi \text{ has a local min at } x \implies \Delta_\infty \varphi(x) \leq 0. \qquad (2.25)$$

In addition, if $D\varphi(x) = 0$, then

$$D^2 \varphi(x) \quad \text{has a nonpositive eigenvalue.} \qquad (2.26)$$

These results follow directly from what was already shown because $-u$ enjoys comparison with cones from above. See Section 3.

2.3 ∞-Harmonic Implies Comparison with Cones

Suppose that $\Delta_\infty u \geq 0$ on the bounded set U. Computing the ∞-Laplacian on a radial function $x \mapsto G(|x|)$ yields

$$\Delta_\infty G(|x|) = G''(|x|)G'(|x|)^2$$

if $x \neq 0$ and from this we find that

$$\Delta_\infty(a|x - z| - \gamma|x - z|^2) = -2\gamma(a - 2\gamma|x - z|)^2 < 0$$

for all $x \in U$, $x \neq z$, if $\gamma > 0$ is small enough. But then if $\Delta_\infty u \geq 0$, $u(x) - (a|x - z| - \gamma|x - z|^2)$ cannot have a local maximum in $V \ll U$ different from z, by the very definition of a viscosity solution of $\Delta_\infty u \geq 0$. Thus if $z \notin V \ll U$ and $x \in V$, we have

$$u(x) - (a|x - z| - \gamma|x - z|^2) \leq \max_{w \in \partial V}(u(w) - (a|w - z| - \gamma|w - z|^2)).$$

Now let $\gamma \downarrow 0$. The full assertions now follow from Section 3.

2.4 Exercises and Examples

Below we explore with curves. A C^1 unit speed curve $\gamma : \mathcal{I} \to U$ on some open interval $\mathcal{I} = (t^-, t^+)$ will be called "maximal" in U if the following holds: (i) if $t^+ < \infty$, then $\lim_{t \uparrow t^+} \gamma(t) =: \gamma(t^+) \in \partial U$ (the limit exists, since γ is unit speed) and (ii) If $-\infty < t^-$, then $\gamma(t^-) \in \partial U$. We also consider the variant where $\mathcal{I} = [0, t^+)$, with the obvious modification.

Exercise 2.4. Let $u \in C^2(U)$ and $Du(x^0) \neq 0$. Let $\gamma \colon \mathcal{I} \to U$, where $\mathcal{I}$ is an open interval in $\mathbb{R}$, be such that $Du(\gamma(t)) \neq 0$. Assume that

$$\dot{\gamma}(t) = \frac{Du(\gamma(t))}{|Du(\gamma(t))|} \quad \text{for } t \in \mathcal{I}. \tag{2.27}$$

Show that

$$\frac{d}{dt} u(\gamma(t)) = |Du(\gamma(t))| \quad \text{and} \quad \frac{d}{dt} |Du(\gamma(t))|^2 = \frac{2}{|Du(\gamma(t))|} \Delta_\infty u(\gamma(t)). \tag{2.28}$$

Conclude that u is ∞-harmonic iff for every $x^0 \in U$ with $Du(x^0) \neq 0$, there is a maximal unit speed curve $\gamma \colon \mathcal{I} \to U$ with the following properties: $0 \in \mathcal{I}$, $\gamma(0) = x^0$, $|Du(\gamma(t))|$ is constant, and $u(\gamma(t)) = u(x^0) + t|Du(x^0)|$.

Exercise 2.5. Show that cone functions are ∞-harmonic on the complement of their vertices and that linear functions are ∞-harmonic. Show that $x \mapsto |x|$ is ∞-subharmonic on $\mathbb{R}^n$.

Exercise 2.6. Show that in $n = 2$ and $u(x,y) = x^{4/3} - y^{4/3}$, then for each $(x_0, y_0) \neq (0,0)$, there is a unit speed C^1 curve $\gamma \colon (-\infty, \infty) \to \mathbb{R}^2$ such that (2.27) holds, $\gamma(0) = (x_0, y_0)$, and $|Du(\gamma(t))| \geq |Du(x_0, y_0)|$.

Exercise 2.7. Let $u \in C^1(U)$ and suppose that for each $x^0 \in U$ with $Du(x^0) \neq 0$ there is a maximal unit speed C^1 curve $\gamma \colon [0, t^+) \to U$ with $\gamma(0) = x^0$ such that $\langle Du(\gamma(t)), \dot{\gamma}(t) \rangle \geq |Du(x^0)|$. Show that u enjoys comparison with cones from above (and so it is ∞-subharmonic). Formulate a similar condition which guarantees that u enjoys comparison with cones from below in U and conclude that the u of Exercise 2.6 is ∞-harmonic. Hints: Suppose $V \ll U$ and $C(x) = a|x - z|$ is a cone function with $z \notin V$. Assuming that $u - C \leq c$ on ∂V, the issue is to show that $u - C \leq c$ in V. If not, there exists $x^0 \in V$ such that

$$u(x^0) - C(x^0) = \max_V (u - C) > c. \tag{2.29}$$

Start the curve γ at x^0 and note $Du(x^0) = DC(x^0)$, so $\langle Du(\gamma(t)), \dot{\gamma}(t) \rangle \geq |a|$. Show that $u(\gamma(t)) - C(\gamma(t))$ is nondecreasing, and conclude that (2.29) cannot hold.

Exercise 2.8. Show that if $u \in C^1(U)$ and satisfies the eikonal equation $|Du| = 1$, then u is ∞-harmonic. Note that this class of functions is not C^2 in general, an example being the distance to an interval on the complement of the interval. Hint: There are a number of ways to do this. For example, one involves showing that $\max_{\overline{B}_r(y)} u = u(y) + r$ and looking ahead to Lemma 4.1, and another (closely related) involves showing that any curve $\gamma(t)$ satisfying (2.27) is a line on which Du is constant.

Exercise 2.9. Let $n = 2, u(x) = x_1, v(x) = |x|$.

(a) Construct an example of a bounded set $U \subset \mathbb{R}^2 \setminus \{0\}$ such that $u < v$ on ∂U except at two points, and $u = v$ holds on the line segment joining these two points (there is no "strong comparison theorem" for ∞-harmonic functions).

(b) Show that the function

$$
w(x) = \begin{cases} u(x) = x_1 \text{ for } x_1 > 0, x_2 > 0, \\ v(x) = |x| \text{ for } x_1 > 0, x_2 \leq 0, \end{cases}
$$

is ∞-harmonic in $\{x_1 > 0\}$ ("unique continuation" does not hold for ∞-harmonic functions). There are a number of ways to do this, including using Exercise 2.8.

Exercise 2.10. The point of (a) of this exercise is to give a proof that if $C(x) = a|x - z|$ is a cone function, then $C \in \mathrm{AML}(\mathbb{R}^n \setminus \{z\})$, which generalizes to the case in which $|\cdot|$ is *any* norm on $\mathbb{R}^n$, and, moreover, if b in (2.4) is $C|_{\partial U}$, then the only solution of (2.1) is C, provided that U is bounded. However, so we don't have to talk too much, assume that $|\cdot|$ is the Euclidean norm here.

(a) Show that $C \in \mathrm{AML}(\mathbb{R}^n \setminus \{z\})$ without using calculus. Hint: Review Section 2.1. Similarly, show that linear functions are in $\mathrm{AML}(\mathbb{R}^n)$.

(b) Let u be a solution of (2.4) where U is bounded, $b = C|_{\partial U}$ and $z \notin U$. Show that for $\varepsilon > 0$ the set $V := \{x \in U : u(x) > C(x) + \varepsilon\}$ is empty. Hint: If it is not, then $u - C = \varepsilon$ on ∂V. Continue to conclude that $u = C$ on U.

(c) Show that if $b(x) = \langle p, x \rangle$ on ∂U in (2.4) and U is bounded, then $u(x) = \langle x, p \rangle$ is the unique solution of (2.4) and it is the maximal ∞-subharmonic function with the property $u = b$ on ∂U.

Remark 2.3. The only comparisons used by Savin in [49] are those of Exercise 2.10.

3 From ∞-Subharmonic to ∞-Superharmonic

We put these short remarks in a section of their own, so that they stand out. The theory we are discussing splits naturally into two halves. Owing to our biases, we present results about ∞-subharmonic functions directly; this is the first half. Then, if u is ∞-superharmonic, the second half is obtained by applying the result for ∞-subharmonic functions to $-u$. In contrast to, say, the Laplace equation $\Delta u = 0$, where one might prove the mean value property without splitting it into halves, we *do not* have this kind of option. The reason lies in the very notion of a viscosity solution (even if this theory applies very well to the Laplace equation). The $u, -u$ game we can play here is a reflection of the definitions or, if you prefer, the fact that $\Delta_\infty(-u) = -\Delta_\infty u$ if u is smooth.

This is true of the other properties we use. For example, u enjoys comparison with cones from above iff $-u$ enjoys comparison with cones from below, and so on.

In this spirit, it is notable that we have not split the "absolutely minimizing" notions into halves, and we will not. However, this can be done. See Section 4 of [8], where further equivalences are given beyond what is discussed in these notes.

The main message is that we often present results or proofs *only* in the ∞-subharmonic case and assume that then the reader knows the corresponding result or proof in the ∞-superharmonic case, and therefore the ∞-harmonic case.

4 More Calculus of ∞-Subharmonic Functions

While we have been assuming that our functions are continuous so far, we will show below that upper-semicontinous functions which enjoy comparison with cones from above are necessarily locally Lipschitz continuous. Thus there is no generality lost in working with continuous functions at the outset (we could have used upper-semicontinuous functions earlier).

The basic calculus type results that we derive about ∞-subharmonic functions below are all consequences of the particular case (2.16) of comparison with cones from above. We have already seen an example of this in Section 2.2. This inequality is equivalent to comparison with cones from above or ∞-subharmonicity by what has already been shown; we also review this in the following proposition. We regard (2.16) as the ∞-subharmonic analogue of the mean value property of ordinary subharmonic functions which estimates the value of such a function at the center of a ball by its average over the ball or the bounding sphere.

The first assertion of the next proposition is that the sphere in (2.16) can be replaced by the ball. The other assertions comprise, together with their consequences in Lemma 4.3, the rest of the most basic facts about ∞-subharmonic functions.

Lemma 4.1. *Let $u\colon U \to \mathbb{R}$ be upper-semicontinuous.*

(a) Assume that

$$u(x) \leq u(y) + \max_{\{w:|w-y|\leq r\}} \left(\frac{u(w) - u(y)}{r} \right) |x - y|. \qquad (4.1)$$

for $y \in U$, $r > 0$, and $x \in \overline{B}_r(y) \subset U$. Then

$$\max_{\{w:|w-y|\leq r\}} u(w) = \max_{\{w:|w-y|=r\}} u(w); \qquad (4.2)$$

in particular, (2.16) holds.

(b) *If u satisfies the conditions of (a), then u is locally Lipschitz continuous in U.*

(c) *If u satisfies the conditions of (a), then u is ∞-subharmonic and enjoys comparison with cones from above.*

(d) *If u satisfies the conditions of (a), then the quantity*

$$S^+(y,r) := \max_{\{w:|w-y|\le r\}} \left(\frac{u(w)-u(y)}{r} \right) = \max_{\{w:|w-y|=r\}} \left(\frac{u(w)-u(y)}{r} \right)$$
(4.3)

is nonnegative and nondecreasing in r, $0 < r <$ dist $(y, \partial U)$. Moreover,

$$\text{if } |w-y| = r \text{ and } S^+(y,r) = \frac{u(w)-u(y)}{r}, \text{ then}$$

$$S^+(y,r) \le S^+(w,s) \text{ for } 0 < s < \text{dist}(y, \partial U) - r.$$
(4.4)

(e) *u satisfies the conditions of (a) if and only if for every $y \in U$*

$$r \mapsto \max_{\overline{B}_r(y)} u$$
(4.5)

is convex on $0 \le r < $ dist $(y, \partial U)$.

Proof. Assume that $y \in U$, (4.1) holds, $\overline{B}_r(y) \subset U$, $|x-y| < r$ and $u(x) = \max_{\overline{B}_r(y)} u$. Then we may replace $u(w)$ by $u(x)$ in (4.1) to conclude that

$$u(x)(1-|x-y|/r) \le u(y)(1-|x-y|/r),$$

which implies that $u(x) \le u(y)$. Since also $u(x) \ge u(y)$, we conclude that $u(x) = u(y)$. Since this is true for all y such that $x \in B_r(y) \ll U$ and $u(x) = \max_{\overline{B}_r(y)} u$, it is true if $\overline{B}_R(x) \subset U$, $u(x) = \max_{\overline{B}_R(x)} u$ and $|y-x| < R/2$. Thus if u has a local maximum point, it is constant in a ball around that point. We record this: if $\overline{B}_R(x) \ll U$,

$$u \text{ satisfies (4.1) and } u(x) = \max_{\overline{B}_R(x)} u, \text{ then } u \text{ is constant on } B_{R/2}(x). \quad (4.6)$$

This guarantees that if u assumes its maximum value at any point of a connected open set, then it is constant in that set, and hence that the maximum of u over any closed ball is attained in the boundary. There is no difference between (4.1) and (2.16) and (a) is proved.

We turn to (b). Assume, to begin, that $u \le 0$. Then, as $u(w) \le 0$ in (4.1), the $u(w)$ on the right can be dropped. Thus we have, written three equivalent ways,

$$\text{(a) } u(x) \le \left(1 - \frac{|x-y|}{r} \right) u(y);$$

$$\text{(b) } -u(y) \le -u(x) \left(\frac{r}{r-|x-y|} \right);$$
(4.7)

$$\text{(c) } u(x) - u(y) \le -\frac{|x-y|}{r} u(y).$$

Either of (4.7) (a) or (b) is a Harnack inequality; (b) is displayed just because you might prefer the nonnegative function $-u$, which is lower-semicontinuous and enjoys comparison with cones from below, to u. If $u(x) \neq 0$, either estimates the ratio $u(y)/u(x)$ by quantities not depending on u. Taking the limit inferior as $y \to x$ on the right of (4.7)(a), we find that u is lower-semicontinuous as well as upper-semicontinuous, so it is continuous. If also $B_r(x) \ll U$, we may interchange x and y in (4.7) (c) and conclude, from the two relations, that

$$|u(x) - u(y)| \leq -\min(u(x), u(y)) \frac{|x - y|}{r - |x - y|}.$$

As u is locally bounded, being continuous, we conclude that it is also locally Lipschitz continuous. If $u \leq 0$ does not hold and $x, y \in B_r(z)$, where $\overline{B}_{2r}(z) \subset U$, replace u by $u - \max_{\overline{B}_{2r}(z)} u$. We thus learn that u is Lipschitz continuous in $B_r(z)$ if $\overline{B}_{2r}(z) \subset U$.

We turn to (c). The assumptions of (a) imply that if $|x - y| = s \leq r$, then

$$\frac{u(x) - u(y)}{s} \leq \max_{\{w:|w-y|=r\}} \left(\frac{u(w) - u(y)}{r} \right).$$

The monotonicity of $S^+(r, y)$ in r follows upon maximizing the left-hand side with respect to x, $|x - y| = s$. The quantity $S^+(y, r)$ is nonnegative by what was already shown - u attains its maximum over a ball on the boundary.

To prove (d), let the assumptions of (4.4) hold: $r < \text{dist}(y, \partial U)$ and

$$|w - y| = r, \quad u(w) = \max_{\overline{B}_r(y)} u. \tag{4.8}$$

Let $0 < s < \text{dist}(y, \partial U)$ and for $0 \leq t \leq 1$ put $y_t := y + t(w - y)$. By our assumptions,

$$u(y_t) - u(y) \leq \left(\frac{u(w) - u(y)}{r} \right) |y_t - y| = t(u(w) - u(y));$$

equivalently,

$$\frac{u(w) - u(y)}{|w - y|} \leq \frac{u(w) - u(y_t)}{|w - y_t|}$$

which implies, using the choice of w and monotonicity of S^+, that

$$S^+(y, r) = \frac{u(w) - u(y)}{|w - y|} \leq \frac{u(w) - u(y_t)}{|w - y_t|} \leq S^+(y_t, s)$$

for $s \geq |w - y_t| = (1 - t)|w - y|$. Letting $t \uparrow 1$ and using the continuity of $S^+(x, s)$ in x, this yields

$$S^+(y, r) \leq S^+(w, s).$$

We turn to (e). If

$$M(r) := \max_{|w-y| \leq r} u(w) \text{ is convex on } 0 \leq r < \text{dist}\,(y, \partial U) \qquad (4.9)$$

we note that then $M(0) = u(y)$ and so

$$\frac{M(s) - M(0)}{s} \leq \frac{M(r) - M(0)}{r} \text{ for } 0 < s \leq r \qquad (4.10)$$

says that

$$u(x) \leq u(y) + \max_{\{w:|w-y|\leq r\}} \left(\frac{u(w) - u(y)}{r} \right) s$$

for $|x - y| \leq s$, and (4.1) holds.

To prove the converse, recall that if (4.1) holds, then so does (2.16). By the the proof of Section 2.2, u is ∞-subharmonic. By Section 2.3, u enjoys comparison with cones from above. Therefore we have the following variant of (2.16):

$$u(x) \leq \left(\max_{|w-y|\leq r} u(w) \right) \frac{|x - y| - s}{r - s} + \left(\max_{|w-y|\leq s} u(w) \right) \frac{r - |x - y|}{r - s} \qquad (4.11)$$

for $0 \leq s \leq |x - y| \leq r$. The inequality is obvious for $|x - y| = s, r$, so it holds as asserted. Taking the maximum of $u(x)$ over $\overline{B}_\tau(y)$, where $s \leq \tau \leq r$ yields

$$\max_{|w-y|\leq\tau} u(w) \leq \left(\max_{|w-y|\leq r} u(w) \right) \frac{\tau - s}{r - s} + \left(\max_{|w-y|\leq s} u(w) \right) \frac{r - \tau}{r - s}, \qquad (4.12)$$

which says that

$$r \mapsto \max_{|w-y|\leq r} u(w) \text{ is convex.} \qquad (4.13)$$

We now know that (2.16) guarantees that u is locally Lipschitz continuous. To use this fact and the estimate implicit in (2.16) efficiently, we introduce the local Lipschitz constant $L(v, x)$ of a function v at a point x.

Definition 4.1. Let $v: U \to \mathbb{R}$ and $x \in U$. Then

$$L(v, x) := \lim_{r\downarrow 0} \text{Lip}(v, B_r(x)) = \inf_{0 < r < \text{dist}\,(x,\partial U)} \text{Lip}(v, B_r(x)). \qquad (4.14)$$

Of course, $L(u, x) = \infty$ is quite possible.

Lemma 4.2. Let $v: U \to \mathbb{R}$.

(a) $L(v, x)$ is upper-semicontinuous in $x \in U$.
(b) If v is differentiable at $y \in U$, then then $L(v, y) \geq |Dv(y)|$.
(c) If $y \in U$ and $L(v, y) = 0$, then v is differentiable at y and $Dv(y) = 0$.

(d) *Let the line segment* $[y, z] \subset U$. *Then*

$$|v(y) - v(z)| \leq \left(\max_{w \in [y,z]} L(v, w) \right) |y - z|.$$

In consequence, $\mathrm{Lip}(v, B_r(y)) \leq \max_{x \in \overline{B}_r(y)} L(v, x)$.

(e) $Dv \in L^\infty(U)$ *holds in the sense of distributions if and only if* $L(v, x)$ *is bounded on* U *and then*

$$\sup_{x \in U} L(v, x) = |||Dv|||_{L^\infty(U)} \text{ and } L(v, x) = \lim_{r \downarrow 0} |||Dv|||_{L^\infty(B_r(x))}. \quad (4.15)$$

Proof. To establish (a), note that if $x^j \to x$, then $B_{r-|x-x^j|}(x^j) \subset B_r(x)$ and therefore, for large j,

$$L(u, x^j) \leq \mathrm{Lip}(u, B_{r-|x-x^j|}(x^j)) \leq \mathrm{Lip}(u, B_r(x))$$

Let $j \to \infty$ and then $r \downarrow 0$ to conclude that $\limsup_{j \to \infty} L(u, x^j) \leq L(u, x)$.

We prove (b) assuming, as we may, that $p := Du(y) \neq 0$, choose $x = y + \lambda p$ with small $\lambda > 0$ to find

$$\mathrm{Lip}(v, B_{\lambda|p|}(y)) \geq \frac{|v(y + \lambda p) - v(y)|}{|\lambda p|} \geq \frac{\lambda |p|^2}{\lambda |p|} + o(1) \text{ as } \lambda \downarrow 0.$$

so, recalling $Dv(y) = p$, $L(v, y) \geq |p|$.

The example $n = 1$, $v(x) = x^2 \sin(1/x)$ and $y = 0$ shows that, in general, $L(v, y) > |Dv(y)|$.

Now assume that $L(v, y) = 0$. Then for each $\varepsilon > 0$ there exists $\delta > 0$ such that

$$|x - y| < \delta \implies |v(x) - v(y)| \leq \mathrm{Lip}(v, B_\delta(y))|x - y| \leq \varepsilon |x - y|,$$

which proves (c).

We turn to (d). For $t \in [0, 1)$, $0 \leq \delta < 1 - t$ and $r > \delta |z - y|$, and

$$g(t) = |v(y + t(z - y)) - v(y)|,$$

we have

$$\frac{g(t + \delta) - g(t)}{\delta} \leq \frac{|v(y + (t + \delta)(z - y)) - v(y + t(z - y))|}{\delta}$$

$$\leq \mathrm{Lip}(v, B_r(y + t(z - y)))|z - y| \text{ for } r \geq \delta |z - y|.$$

Letting $\delta \downarrow 0$ yields

$$\limsup_{\delta \downarrow 0} \frac{g(t + \delta) - g(t)}{\delta} \leq \mathrm{Lip}(v, B_r(y + t(z - y)))|z - y|$$

and then, letting $r \downarrow 0$,

$$\limsup_{\delta \downarrow 0} \frac{g(t+\delta) - g(t)}{\delta} \le L(v, y + t(z-y))|z-y| \le \max_{w \in [y,z]} L(v,w)|z-y|.$$

This implies, using elementary facts about Dini derivatives (for example), that

$$g(1) - g(0) = |v(z) - v(y)| \le \max_{w \in [y,z]} L(v,w)|z-y|,$$

as claimed.

The claim (e) follows easily from (d) if we take as known that $Dv \in L^\infty(U)$ and $\||Dv|\|_{L^\infty(U)} = L$ if and only if $[y,z] \subset U$ implies

$$|v(z) - v(y)| \le L|z-y|$$

and L is the least such constant.

Definition 4.2. *If u is ∞-subharmonic in U and $x \in U$, then*

$$S^+(x) := \lim_{r \downarrow 0} S^+(x,r) = \lim_{r \downarrow 0} \max_{\{w : |w-x|=r\}} \left(\frac{u(w) - u(x)}{r} \right). \tag{4.16}$$

Similarly, if u is ∞-superharmonic, then

$$S^-(x) := \lim_{r \downarrow 0} S^-(x,r) := \lim_{r \downarrow 0} \min_{\{w : |w-x|=r\}} \left(\frac{u(w) - u(x)}{r} \right). \tag{4.17}$$

Remark 4.1. Note the notational peculiarities. S^+ is used both with two arguments and with a single argument; no confusion should arise from this once noted, but the meaning changes with the usage. L has two arguments, and, in contrast to S^+, one of them is the function itself. (We did not want to write $S^+(u,x,r)$ or some variant.) We will want to display the function argument later, and the identities (4.18), (4.19) below will allow us to use $L(u,x)$ in place of $S^+(x), S^-(x)$ when we need to indicate the function involved.

Lemma 4.3. *Let $u \in C(U)$ be ∞-subharmonic. Then for $x \in U$ and $B_r(x) \ll U$,*

$$L(u,x) = S^+(x). \tag{4.18}$$

In consequence, if u is ∞-harmonic in U and $x \in U$, then

$$S^-(x) = -S^+(x). \tag{4.19}$$

Moreover, $u \in C(U)$ satisfies

$$L(u,x) \le \max_{\{w : |w-x| \le r\}} \left(\frac{u(w) - u(x)}{r} \right) \tag{4.20}$$

for $\overline{B}_r(x) \subset U$ if and only if u is ∞-subharmonic in U. Finally, at any point y of differentiability of u,

$$|Du(y)| = S^+(y) = L(u,y). \tag{4.21}$$

Proof. First, via Lemma 4.3, we have that $\max_{\overline{B}_r(x)} L(u,x) = \mathrm{Lip}(u,\overline{B}_r(x))$. Thus

$$S^+(x,r) = \max_{\{w:|w-x|\leq r\}} \left(\frac{u(w) - u(x)}{r} \right) \leq \max_{\{w:|w-x|\leq r\}} L(u,w).$$

The upper-semicontinuity of $L(u,x)$ in x (Lemma 4.2 (a)) then implies, upon taking the limit $r \downarrow 0$, that $S^+(x) \leq L(u,x)$. To obtain the other inequality, let $[w,z] \subset U$. By the local Lipschitz continuity of u, $g(t) := u(w + t(z - w))$ is Lipschitz continuous in $t \in [0,1]$. Fix $t \in (0,1)$ and observe that the definition of S^+ implies, for small $h > 0$,

$$\frac{g(t+h) - g(t)}{h} = \frac{u(w + (t+h)(z-w)) - u(w + t(z-w))}{h|z-w|}|z-w|.$$

$$\leq S^+(w + t(z-w), h|z-w|)|z-w|.$$

The inequality follows from the definition of $S^+(w+t(z-w),r)$. Sending $h \downarrow 0$ we find

$$\limsup_{h\downarrow 0} \frac{g(t+h) - g(t)}{h} \leq S^+(w + t(z-w))|z-w|$$

$$\leq \left(\sup_{y\in[w,z]} S^+(y) \right)|z-w|.$$

Thus

$$u(z) - u(w) = g(1) - g(0) \leq \left(\sup_{y\in[w,z]} S^+(y) \right)|z-w|.$$

Interchanging z and w we arrive at

$$|u(z) - u(w)| \leq \left(\sup_{y\in[w,z]} S^+(y) \right)|z-w|. \tag{4.22}$$

Using also the monotonicity of $S^+(x,r)$ in r (Lemma 4.1), if $\delta > 0$ is small, we have, via (4.22),

$$\mathrm{Lip}(u, B_r(x^0)) \leq \sup_{x\in\overline{B}_r(x^0)} S^+(x) \leq \sup_{x\in B_r(x^0)} S^+(x,\delta).$$

Now send $r \downarrow 0$ and then $\delta \downarrow 0$ to find, via the continuity of $S^+(x,\delta)$ for $\delta > 0$,

$$L(u,x^0) \leq S^+(x^0,\delta) \downarrow S^+(x^0) \quad \text{as } \delta \downarrow 0;$$

this completes the proof of (4.18).

To establish (4.19), note that $L(u,x) = L(-u,x)$ and that S^- for u is just $-S^+$ for $-u$ and invoke (4.18).

If u is differentiable at y and $|w - y| = r$, then

$$u(w) = u(y) + \langle Du(y), w - y \rangle + o(r) \implies$$

$$\max_{\{w : |w-y| = r\}} \left(\frac{u(w) - u(y)}{r} \right) = |Du(y)| + \frac{o(r)}{r}.$$

Now use (4.18) to conclude that (4.21) holds.

We next show that (4.20) implies (4.1), and thus that u is ∞-subharmonic. Suppose that $t \to \gamma(t)$ is a C^1 curve in U. Using (4.20) with $\gamma(t)$ in place of x, one easily checks, using the local Lipschitz continuity of $t \mapsto u(\gamma(t))$, that, almost everywhere,

$$\left| \frac{d}{dt} u(\gamma(t)) \right| \leq L(u, \gamma(t)) |\dot{\gamma}(t)|$$

$$\leq \max_{w \in \overline{B}_r(\gamma(t))} \left(\frac{u(w) - u(\gamma(t))}{r} \right) |\dot{\gamma}(t)| \text{ for } r < \text{dist}\,(\gamma(t), \partial U). \tag{4.23}$$

It is convenient to rewrite (4.23) as

$$\pm \frac{d}{dt} u(\gamma(t)) + \frac{|\dot{\gamma}(t)|}{r} u(\gamma(t)) \leq \left(\max_{w \in \overline{B}_r(\gamma(t))} u(w) \right) \frac{|\dot{\gamma}(t)|}{r} \text{ for } r < \text{dist}\,(\gamma(t), \partial U). \tag{4.24}$$

If $x \in B_r(y) \ll U$ and $\gamma(t) = y + t(x - y)$, then $\text{dist}\,(\gamma(t), \partial U) > r - t|x - y|$. Moreover, $B_r(y) \supset B_{r-t|x-y|}(\gamma(t))$. We use this information in (4.24) to deduce that

$$\pm \frac{d}{dt} u(\gamma(t)) + \frac{|x - y|}{r - t|x - y|} u(\gamma(t)) \leq \left(\max_{|w-y| \leq r} u(w) \right) \frac{|x - y|}{r - t|x - y|}. \tag{4.25}$$

This simple differential inequality taken with the "+" sign and integrated over $0 \leq t \leq 1$ yields (4.1).

Note that the generality of the "$\pm$" above is superfluous; it corresponds to reversing the direction of γ.

Exercise 4.1. Perform the integration of (4.25) referred to just above.

Exercise 4.2. Show that the map $C(U) \ni u \mapsto L(u, x)$ ($x \in U$ is fixed) is lower-semicontinuous but it is not continuous. Show, however, that the restriction of this mapping to the set of ∞-subharmonic functions u is continuous and $(u, x) \mapsto L(u, x)$ is upper-semicontinuous.

Exercise 4.3. If u is ∞-subharmonic in U and $u \leq 0$, use (4.20) to conclude that

$$|Du(x)| \leq -\frac{u(x)}{\text{dist}\,(x, \partial U)} \text{ if } u \text{ is differentiable at } x \in U.$$

Exercise 4.4. Show that (4.20) always holds with equality for cone functions $C(x) = a|x - z|$ with nonnegative slopes, so C is ∞-subharmonic on $\mathbb{R}^n$ and (4.20) is sharp. Observe that (4.20) fails to hold for cones with negative slopes.

Exercise 4.5. Let u be ∞-subharmonic in U and $u \leq 0$. It then follows from (4.24) that

$$\frac{d}{dt} u(\gamma(t)) + \frac{|\dot{\gamma}(t)|}{\text{dist}\,(\gamma(t), \partial U)} u(\gamma(t)) \leq 0. \tag{4.26}$$

(i) If $\gamma \colon [0,1] \to U$, conclude that

$$u(\gamma(1)) \leq \exp\left(-\int_0^1 \frac{|\dot{\gamma}(t)|}{\text{dist}\,(\gamma(t), \partial U)}\, dt\right) u(\gamma(0)). \tag{4.27}$$

(ii) Use (i) to show that if $x, y \in B_r(z) \subset B_R(z) \subset U$, then

$$e^{\frac{|x-y|}{R-r}} u(x) \leq u(y). \tag{4.28}$$

This is another Harnack inequality.

(iii) Let $\mathcal{H} = \{(x_1, \ldots, x_n) : x_n > 0\}$ be a half-space, $\Delta_\infty u \geq 0$ and $u \leq 0$ in $\mathcal{H}$. Using (4.26), show that $x_n \mapsto x_n u(x_1, \ldots, x_n)$ is nonincreasing and $x_n \mapsto u(x_1, \ldots, x_n)/x_n$ is nondecreasing on $\mathcal{H}$. (Here one only needs to check the sign of a derivative.)

(iv) Let $u \leq 0$ be ∞-subharmonic in $B_R(0)$. Use (4.26) to show that if $x \in \mathbb{R}^n$, then

$$t \mapsto \frac{u(tx)}{R - t|x|} \text{ is nonincreasing on } 0 \leq t|x| < R.$$

In particular, if $x \in B_R(0)$, then $u(x)/(R - |x|) \leq u(0)/R$.

Remark 4.2. See Sections 8 and 9 concerning citations for Exercises 4.3 and 4.5. Clearly, in Exercise 4.5 we are showing how to organize various things in the literature in an new efficient way.

5 Existence and Uniqueness

We have nothing new to offer regarding existence, and as far as we know Theorem 3.1 of [8] remains the state of the art in this regard. The result allows U to be unbounded and a boundary function b which grows at most linearly.

Theorem 5.1. *Let U be an open subset of $\mathbb{R}^n$, $0 \in \partial U$, and $b \in C(\partial U)$. Let $A^\pm, B^\pm \in \mathbb{R}$, $A^+ \geq A^-$, and*

$$A^-|x| + B^- \leq b(x) \leq A^+|x| + B^+ \text{ for } x \in \partial U. \tag{5.1}$$

Then there exists $u \in C(\overline{U})$ which is ∞-harmonic in U, satisfies $u = b$ on ∂U and which further satisfies

$$A^-|x| + B^- \leq u(x) \leq A^+|x| + B^+ \quad \text{for } x \in \overline{U}. \tag{5.2}$$

The proof consists of a rather straight-foward application of the Perron method, using the equivalence between ∞-harmonic and comparison with cones. In the statement, the assumption $0 \in \partial U$ is made to simplify notation and interacts with the assumption (5.1). A translation handles the general case, but this requires naming a point of ∂U. The Perron method runs by defining $\underline{h}, \overline{h} : \mathbb{R}^n \to \mathbb{R}$ by

$$\underline{h}(x) = \sup\{\underline{C}(x) : \underline{C}(x) = a|x - z| + c, \ a < A^-, c \in \mathbb{R}, z \in \partial U, \ \underline{C} \leq b \text{ on } \partial U\},$$

$$\overline{h}(x) = \inf\{\overline{C}(x) : \overline{C}(x) = a|x - z| + c, \ a > A^+, c \in \mathbb{R}, z \in \partial U, \ \overline{C} \geq b \text{ on } \partial U\}.$$

and then showing that u below has the desired properties.

$$u(x) := \sup\{v(x) : \underline{h} \leq v \leq \overline{h} \text{ and } v \text{ enjoys comparison with cones from above}\}.$$
$$(5.3)$$

See [8] for details, or give the proof as an exercise.

Uniqueness has always been a sore spot for the theory, in the sense that it took a long time for Jensen [36] to give the first, quite tricky, proof and then another proof, still tricky, but more in line with standard viscosity solution theory, was given by Barles and Busca [9]. A self-contained presentation of the proof of [9], which does not require familiarity with viscosity solution theory, is given in [8].

Here we give the skeleton of a third proof, from [28], in which unbounded domains are treated for the first time. This time, however, we fully reduce the result to standard arguments from viscosity solution theory, and we do not render our discussion self-contained in this regard.

The result, proved first by Jensen and with a second proof by Barles and Busca, is

Theorem 5.2. *Let U be bounded, $u, v \in C(\overline{U})$, $\Delta_\infty u \geq 0$ and $\Delta_\infty v \leq 0$ in U. Then if $u \leq v$ on ∂U, we have $u \leq v$ in U.*

We just sketch the new proof, as it applies to the case of a bounded domain. The main point is an approximation result.

Proposition 5.1. *Let U be a (possibly unbounded) subset of $\mathbb{R}^n$ and $u \in C(\overline{U})$ be ∞-subharmonic. Let $\varepsilon > 0$,*

$$U_\varepsilon := \{x \in U : L(u, x) \geq \varepsilon\} \ \text{ and } \ V_\varepsilon := \{x \in U : L(u, x) < \varepsilon\}. \tag{5.4}$$

Then there is function u_ε with the properties (a)-(f) below:

(a) $u_\varepsilon \in C(\overline{U})$ and $u_\varepsilon = u$ on ∂U.
(b) $u_\varepsilon = u$ on U_ε.
(c) $\varepsilon - |Du_\varepsilon| = 0$ (in the viscosity sense) on V_ε.
(d) $L(u_\varepsilon, x) \geq \varepsilon$ for $x \in U$.
(e) u_ε is ∞-subharmonic in U.
(f) $u_\varepsilon \leq u$ and $\lim_{\varepsilon \downarrow 0} u_\varepsilon = u$.

Remark 5.1. In (c), "viscosity sense" means that at a maximum (minimum) $\hat{x}$ of $u_\varepsilon - \varphi$ we have $\varepsilon - |D\varphi(\hat{x})| \leq 0$ $(\varepsilon - |D\varphi(\hat{x})| \geq 0)$. For example, $-\varepsilon|x|$ has this property on $\mathbb{R}^n$, while $\varepsilon|x|$ does not. This is consistent with the conventions of [31].

The assertions (a)-(c) are really a prescription of how to construct u_ε, which is then given by a standard formula. The trickiest point is (e), which we discuss below. Full details are available in [28]; see also Barron and Jensen [13], Proposition 5.1, where related observations were first made, although with a different set of details; one does not find Proposition 5.1 in [13], but there is an embedded proof of (e), different from the one we will sketch. In fact, these authors show that if u is ∞-harmonic, then the u_ε above solves $\max(-\Delta_\infty u_\varepsilon, \varepsilon - |Du_\varepsilon|) = 0$; this variational inequality played a fundamental role in [36] (see Section 8). However, they do not consider subsolutions nor note the approximation property (f) in this case.

We first explain how Proposition 5.1 reduces Theorem 5.2 to a routine citation of results in [31]. After that, we present our proof of the most subtle point, (e), of Proposition 5.1; that discussion will also verify (d).

If we can show that

$$u_\varepsilon(x) - v(x) \leq \max_{\partial U}(u_\varepsilon - v) \text{ for } x \in U \tag{5.5}$$

then (f) allows us to replace u_ε by u and the right and Theorem 5.2 follows in the limit $\varepsilon \downarrow 0$. Thus, using (d), we may assume, without further ado, that

$$L(u, x) \geq \varepsilon \text{ for } x \in U. \tag{5.6}$$

Some version of the following, which is the step that allows us to take advantage of (5.6), is used in all comparison proofs for Δ_∞ below, beginning with the proof of Jensen [36]. This sort of nonlinear change of variables is a standard tool in comparison theory of viscosity solutions. Let $\lambda > 0$ and $\sup_U u < \frac{1}{2\lambda}$ and then define w by

$$u(x) = G(w(x)) := w(x) - \frac{\lambda}{2}w(x)^2 \text{ and } w(x) < \frac{1}{\lambda}. \tag{5.7}$$

That is, $w(x) = (1 - \sqrt{1 - 2\lambda u})/\lambda$. Clearly $w \to u$ uniformly as $\lambda \downarrow 0$. It therefore suffices to show that

$$w(x) - v(x) \leq \max_{\partial U}(w - v) \text{ for } x \in U \tag{5.8}$$

To this end, one formally computes that

$$\begin{aligned}
0 \leq \Delta_\infty u &= \langle D^2 u Du, Du \rangle \\
&= \langle (G''(w)(Dw \otimes Dw) + G'(w)D^2 w) G'(w)Dw, G'(w)Dw \rangle \\
&= G''(w)G'(w)^2 |Dw|^4 + G'(w)^3 \langle D^2 w Dw, Dw \rangle,
\end{aligned} \tag{5.9}$$

which implies, using (5.6), that

$$\Delta_\infty w = \langle D^2 w Dw, Dw \rangle \geq -\frac{G''(w)}{G'(w)}|Dw|^4$$

$$= -\frac{G''(w)}{G'(w)^3}|Du|^4 \geq \frac{\lambda}{(1-\lambda w)^3}\varepsilon^4 \geq \kappa > 0 \tag{5.10}$$

for some κ. It is straightforward to check that these computations are valid in the viscosity sense. It follows immediately from Theorem 3.1 in [31], used in the standard way, that $\Delta_\infty w \geq \kappa > 0$ and $\Delta_\infty v \leq 0$ together imply that $w - v$ cannot have an interior local maximum, whence (5.8).

We now discuss key elements of the proof of Proposition 5.1. Note that V_ε is open by the upper-semicontinuity of $L(u, \cdot)$ while $U_\varepsilon = U \setminus V_\varepsilon$ is closed relative to U.

It is known that any function solving $\varepsilon - |Du_\varepsilon| = 0$ in the viscosity sense in an open set is ∞-subharmonic in that set. In fact, from the general theory of viscosity solutions of first order equations, one has the formula

$$u_\varepsilon(x) = \max_{\{y:|y-z|=r\}} (u_\varepsilon(y) - \varepsilon|x - y|) \text{ for } x \in \overline{B}_r(z), \tag{5.11}$$

whenever $\overline{B}_r(z) \subset V_\varepsilon$, which makes the claim evident (in our context, we could use that the class of functions enjoying comparison with cones from above is closed under taking supremums in place of the analogous statement about viscosity subsolutions). The function u_ε is produced by a similar standard formula, with ∂V_ε in place of $\partial B_r(z)$ and the distance interior to V_ε in place of $|x - y|$ (which is the distance from x to y interior to $\overline{B}_r(z)$).

In the end, given the tools in hand, the main point one needs to establish in order to prove (e) is this: assuming that we have constructed u_ε, then

$$x \in (\partial V_\varepsilon) \cap U \implies L(u_\varepsilon, x) \leq L(u, x). \tag{5.12}$$

Since $x \in U_\varepsilon \supset (\partial V_\varepsilon) \cap U$, we have $L(u, x) \geq \varepsilon$, which renders the ε's appearing in estimates appearing below harmless as regards establishing (5.12).

To establish (5.12), it suffices to show that if $[y, z] \subset B_r(x) \ll U$, then

$$|u_\varepsilon(z) - u_\varepsilon(y)| \leq \text{Lip}(u, B_r(x))|y - z|, \tag{5.13}$$

for then $\text{Lip}(u_\varepsilon, B_r(x)) \leq \text{Lip}(u, B_r(x))$.

First note that if *any* open interval $(y, z) \subset V_\varepsilon$, then $|Du_\varepsilon| \leq \varepsilon$ a.e. on the open set V_ε implies

$$|u_\varepsilon(z) - u_\varepsilon(y)| \leq \varepsilon|y - z|. \tag{5.14}$$

In particular, if $[y, z] \subset V_\varepsilon$, then (5.13) holds. If $[y, z] \cap U_\varepsilon$ is not empty, we proceed as follows. The set U_ε is closed in U, so $[y, z] \cap U_\varepsilon$ is closed and there exists a least $t_0 \in [0, 1]$ and a greatest $t_1 \in [0, 1]$ such that

$$x^0 := y + t_0(w - y) \in U_\varepsilon, \quad x^1 := y + t_1(w - y) \in U_\varepsilon. \tag{5.15}$$

Then

$$
\begin{aligned}
|u_\varepsilon(z) - u_\varepsilon(y)| &= |u_\varepsilon(z) - u_\varepsilon(x^1) + u_\varepsilon(x^1) - u_\varepsilon(x^0) + u_\varepsilon(x^0) - u_\varepsilon(y)| \\
&= |u_\varepsilon(z) - u_\varepsilon(x^1) + u(x^1) - u(x^0) + u_\varepsilon(x^0) - u_\varepsilon(y)| \\
&\leq |u_\varepsilon(z) - u_\varepsilon(x^1)| + |u(x^1) - u(x^0)| + |u_\varepsilon(x^0) - u_\varepsilon(y)| \\
&\leq |u_\varepsilon(z) - u_\varepsilon(x^1)| + \mathrm{Lip}(u, B_r(x))|x^1 - x^0| + |u_\varepsilon(x^0) - u_\varepsilon(y)|
\end{aligned}
\tag{5.16}
$$

Now each interval $[y, x^0)$ and $(x^1, z]$ is either empty (as is the case, for example, for $[y, x^0)$ if $y = x^0$) or lies entirely in V_ε. Using (5.14) twice, with $z = x^0$ and then with $y = x^1$,

$$
|u_\varepsilon(z) - u_\varepsilon(x^1)| \leq \varepsilon|z - x^1|, \quad |u_\varepsilon(x^0) - u_\varepsilon(y)| \leq \varepsilon|x^0 - y|.
$$

Combining this with (5.16), (5.13) follows from

$$
|z - y| = |z - x^1| + |x^1 - x^0| + |x^0 - y|.
$$

We continue. Let u_ε be as in the proposition and $x \in U$. According to Lemma 4.3, if we can show that

$$
L(u_\varepsilon, x) \leq \max_{\{w : |w - z| \leq r\}} \left(\frac{u_\varepsilon(w) - u(x)}{r} \right)
\tag{5.17}
$$

for $r < \mathrm{dist}\,(x, \partial U)$, we are done. If $x \in U_\varepsilon$, then (5.17) holds. Indeed, then, using (5.12) if necessary (i.e, if $x \in \partial V_\varepsilon$), we have

$$
L(u_\varepsilon, x) \leq L(u, x) \leq \frac{u(w_r) - u(x)}{r} = \frac{u_\varepsilon(w_r) - u_\varepsilon(x)}{r}
\tag{5.18}
$$

for some w_r, $|w_r - x| = r$, which satisfies $L(u, w_r) \geq L(u, x) \geq \varepsilon$, so $w_r \notin V_\varepsilon$ and so $u(w_r) = u_\varepsilon(w_r)$. Hence (5.17) holds if $x \in U_\varepsilon$.

To handle the case $x \in V_\varepsilon$, one first observes that (5.11) implies that if $B_r(z) \subset V_\varepsilon$ then

$$
\max_{y \in \overline{B}_r(z)} u_\varepsilon(y) = u_\varepsilon(z) + \varepsilon r,
\tag{5.19}
$$

and $L(u_\varepsilon, z) = \varepsilon$. Recalling $L(u, x) \geq \varepsilon$ for $x \in U_\varepsilon$, between (5.18) and (5.19), we learn that for every $z \in U$ there is an $r_z > 0$ such that

$$
\max_{w \in \overline{B}_r(z)} u_\varepsilon(w) \geq u_\varepsilon(z) + \varepsilon r \text{ for } 0 \leq r \leq r_z.
$$

This implies, with a little continuation argument, that

$$
\varepsilon \leq \max_{\{w : |w - z| \leq r\}} \left(\frac{u_\varepsilon(w) - u_\varepsilon(x)}{r} \right) \text{ for } r < \mathrm{dist}\,(x, \partial U),
$$

and we are done.

Exercise 5.1. Provide the little continuation argument.

6 The Gradient Flow and the Variational Problem for $|||Du|||_{L^\infty}$

Let us note right away that if u is absolutely minimizing for $\mathcal{F}_\infty$, then it is absolutely minimizing for Lip.

Proposition 6.1. *Let* $u \in C(U)$ *be absolutely minimizing for* $\mathcal{F}_\infty$, *that is, whenever* $V \ll U$, $v \in C(\overline{V})$ *and* $u = v$ *on* ∂V, *then* $\mathcal{F}_\infty(u, V) \leq \mathcal{F}_\infty(v, V)$. *Then* u *is absolutely minimizing for* Lip *(and hence ∞-harmonic).*

Proof. One proof was already indicated in Exercise 2.3. Here is another. Let $v \in C(\overline{V})$ and $u = v$ on ∂V. Assume $\mathrm{Lip}(v, \partial V) < \infty$ and replace v by $\mathcal{M}W_*(v|_{\partial V})$ so that we may assume that $\mathrm{Lip}(v, V) = \mathrm{Lip}(v, \partial V)$. Then, by assumption, $\mathcal{F}_\infty(u, V) \leq \mathcal{F}_\infty(v, V)$, which is at most $\mathrm{Lip}(v, V)$. Now use Exercise 6.1 below.

Exercise 6.1. Show that if $u \in C(\overline{V})$ and $\mathcal{F}_\infty(u, V) \leq \mathrm{Lip}(u, \partial V)$, then $\mathrm{Lip}(u, V) = \mathrm{Lip}(u, \partial V)$.

The main result of this section is the tool we will use to prove the converse to the proposition above.

Proposition 6.2. *Let* u *be* ∞-subharmonic *in* U, $x \in U$. *Then there is a* $T > 0$ *and Lipschitz continuous curve* $\gamma \colon [0, T) \to U$ *with the following properties:*

$$
\begin{aligned}
&\text{(i) } \gamma(0) = x \\
&\text{(ii) } |\dot\gamma(t)| \leq 1 \ a.e. \ on \ [0, T). \\
&\text{(iii) } L(u, \gamma(t)) \geq L(u, x) \ on \ [0, T). \\
&\text{(iv) } u(\gamma(t)) \geq u(x) + tL(u, x) \ on \ [0, T). \\
&\text{(v) } t \mapsto u(\gamma(t)) \ is \ convex \ on \ [0, T). \\
&\text{(vi) } Either \ T = \infty \ or \lim_{t \uparrow T} \gamma(t) \in \partial U.
\end{aligned} \tag{6.1}
$$

The motivation for this result is contained in Exercises 2.4–2.7. We are able, in the general case, to obtain curves with similar properties to those discussed in the exercises.

Before proving this result, let us give an application.

Theorem 6.1. *Let* $u \in \mathrm{AML}(U)$. *Let* $V \ll U$, $v \in C(\overline{V})$ *and* $u = v$ *on* ∂V. *Then* $\sup_V L(u, x) \leq \sup_V L(v, x)$. *In other words, if* u *is* ∞-harmonic, then *it is absolutely minimizing for* $\mathcal{F}_\infty$.

Proof. Under the assumptions of the theorem, if the conclusion does not hold, then there exists $z \in V$ and $\delta > 0$ such that $L(u, z) \geq L(v, x) + \delta$ for $x \in V$. Let $\gamma \colon [0, T) \to V$ be the curve provided by Proposition 6.2 which starts at z (regarding V as the U of the proposition). Since u is bounded on $\overline{V}$, it follows from (iv), (vi) of the proposition that $T < \infty$ and $\lim_{t \uparrow T} \gamma(t) =: \gamma(T) \in \partial V$.

By (iii) of the proposition, $L(u, \gamma(t)) \geq L(u, z) > \sup_V L(v, x)$ and so, using (ii), almost everywhere,

$$\frac{d}{dt} v(\gamma(t)) \leq L(v, \gamma(t)) < L(u, z).$$

Integrating over $[0, T]$ and using (ii) of the proposition again,

$$v(\gamma(T)) - v(z) < TL(u, z) \leq u(\gamma(T)) - u(z).$$

Now $-u$ is also ∞-subharmonic and it is related to $-v$ as u was to v, so there is another curve $\tilde{\gamma}(t) \colon [0, \tilde{T}) \to V$ with $\tilde{\gamma}(\tilde{T}) \in \partial V$ such that

$$-v(\tilde{\gamma}(\tilde{T})) + v(z) < \tilde{T}L(u, z) \leq -u(\tilde{\gamma}(\tilde{T})) + u(z).$$

Adding the two inequalities yields

$$v(\gamma(T)) - v(\tilde{\gamma}(\tilde{T})) < u(\gamma(T)) - u(\tilde{\gamma}(\tilde{T})),$$

which contradicts $u = v$ on ∂U.

Proof of Proposition 6.2

The main idea is to build a discrete version of the desired γ by using the "increasing slope estimate," Lemma 4.1 (d). We may assume that $L(u, x) > 0$, for otherwise we may take $\gamma(t) \equiv x$. Fix δ, $0 < \delta < \text{dist}(x^0, \partial U)$. Form a sequence $\left\{x_\delta^j\right\}_{j=0}^J \subset U$ is according to $x_\delta^0 := x$ and

$$|x_\delta^{j+1} - x_\delta^j| = \delta, \ u(x_\delta^{j+1}) = \max_{\overline{B}_\delta(x_\delta^j)} u \text{ for } j = 1, \ldots, J-1. \qquad (6.2)$$

We allow J to be finite or infinite. The value of J is determined by checking, after the successful determination of some x^j, if $\overline{B}_\delta(x_\delta^j) \subset U$ or not. If so, then x_δ^{j+1} is determined and $j + 1 \leq J$. If not, then $J = j$, and we stop. Clearly, J is at least the greatest integer less than $\text{dist}(x^0, \partial U)/\delta$. According to the increasing slope estimate (4.4), we then have

$$S^+(x_\delta^{j+1}) \geq \frac{u(x_\delta^{j+1}) - u(x_\delta^j)}{\delta} = S^+(x_\delta^j, \delta) \geq S^+(x_\delta^j); \qquad (6.3)$$

Thus $S^+(x_\delta^j) \geq S^+(x_\delta^0) = S^+(x)$ for $j = 0, 1, \ldots J - 1$ and (recall $x_\delta^0 = x$)

$$u(x_\delta^{j+1}) - u(x_\delta^j) \geq \delta S^+(x_\delta^0) \implies u(x_\delta^j) - u(x) \geq j\delta S^+(x). \qquad (6.4)$$

Form the piecewise linear curve defined by $\gamma_\delta(0) = x$ and

$$\gamma_\delta(t) = x_\delta^j + (t - j\delta)\left(\frac{x^{j+1} - x^j}{\delta}\right) \text{ on } j\delta \leq t \leq (j+1)\delta, \ j = 0, \ldots, J-1.$$

For $j = 0, 1, \ldots, J$ we have, by the construction,

$$
\begin{aligned}
&\text{(i) } \gamma_\delta(0) = x \\
&\text{(ii) } |\dot{\gamma}_\delta(t)| = 1 \text{ a.e. on } [0, J\delta). \\
&\text{(iii) } L(u, \gamma_\delta(j\delta)) \geq L(u, x). \\
&\text{(iv) } u(\gamma_\delta(j\delta)) \geq u(x) + j\delta L(u, x).
\end{aligned}
\tag{6.5}
$$

By construction, $\delta J \geq \text{dist}(x, \partial U) - \delta$. By compactness, there is then a sequence $\delta_k \downarrow 0$ and a $\gamma : [0, \text{dist}(x, \partial U)) \to U$ such that $\gamma_{\delta_k}(t) \to \gamma(t)$ uniformly on compact subsets of $[0, \text{dist}(x, \partial U))$. Clearly $\text{dist}(\gamma(t), \partial U) \geq \text{dist}(x, \partial U) - t$. Moreover, if $0 \leq t < \text{dist}(u, \partial U)$, there exist j_k such that $j_k \delta_k \to t$. Passing to the limit in the relations (6.5) with $\delta = \delta_k$ and $j = j_k$, using the upper-semicontinuity of $L(u, \cdot)$, yields all of the relations of (6.1) except (v), (vi). To see that (v) holds, note that the piecewise linear function $g_k(t)$ whose value at $j\delta_k$ is $u(\gamma_{\delta_k}(j\delta_k))$ is convex by (6.3). Moreover, by the continuity of u and the uniform convergence of γ_{δ_k} to γ, g_k converges to $u(\gamma(t))$, which is therefore convex.

The property (vi) can now be obtained by the standard continuation argument of ordinary differential equations. There is a curve γ with the properties of (6.1) defined on a maximal interval of existence of the form $[0, T)$. Assume now that $T < \infty$ and $\lim_{t \uparrow T} \gamma(t) =: \gamma(T)$ and $\gamma(T) \notin \partial U$. The proof concludes by arguing that then γ was not maximal. Indeed, clearly

$$
\lim_{t \uparrow T} \frac{u(\gamma(T)) - u(\gamma(t))}{T - t} \leq L(u, \gamma(T)),
\tag{6.6}
$$

so if we use the construction above, starting at $\gamma(T)$, to extend γ to a curve $\tilde{\gamma} : [0, T + \text{dist}(\gamma(T), \partial U)) \to U$, we obtain a strict extension of γ with all the right properties, producing a contradiction. The property (6.6), together with $u(\tilde{\gamma}(t)) \geq (t - T)L(u, \gamma(T)) + u(\gamma(T))$ for $t \geq T$ and the convexity of $u(\tilde{\gamma}(t))$ on $T \leq t < T + \text{dist}(\gamma(T), \partial U)$ is what guarantees that $u(\tilde{\gamma}(t))$ is convex; it all glues together right, just as in ode. □

Remark 6.1. All of the conclusions which can be obtained using the curves γ of Proposition 6.2 can also be obtained using the discrete ingredients from which they are built, extended "maximally." See [29], [8]. However, it is considerably more elegant to use curves instead of the discrete versions. Barron and Jensen first constructed analogous curves in [13]; however, the technical surroundings in [13] are a bit discouraging from the point of view of extracting this information. In addition, they do not pay attention to the maximality and there is a oversight in their proof of (v) (we learned to include (v) in the list of properties from [13]).

Exercise 6.2. Let $u \in \text{AML}(U) \cap C(\overline{U})$ and U be bounded. Show that $\text{Lip}(u, U) = \text{Lip}(u, \partial U)$. Hint: Exercise 6.1.

Exercise 6.3. Let $u, v \in C(\overline{U})$ and U be bounded. If u is ∞-harmonic in U and $u = v$ on ∂U, show that $\mathcal{F}_\infty(u, U) \leq \mathcal{F}_\infty(v, U)$. That is, u solves the minimum problem for $\mathcal{F}_\infty$ with $b = u|_{\partial U}$.

7 Linear on All Scales

7.1 Blow Ups and Blow Downs are Tight on a Line

The information in this section remains the main evidence we have regarding the primary open problem in the subject: are ∞-harmonic functions C^1? Savin has proved that they are in the case $n = 2$, using the information herein, in the form of Exercise 7.3, which is a version of Proposition 7.1 (a). His paper also requires Proposition 7.1 (b), which is original here, in one argument in which (a) is erroneously cited in support of the argument.

Let $\Delta_\infty u \leq 0$ in U, $x^0 \in U$ and $r > 0$. Then

$$v_r(x) := \frac{u(rx + x^0) - u(x^0)}{r} \tag{7.1}$$

satisfies

$$\mathrm{Lip}(v_r, B_R(0)) = \mathrm{Lip}(u, B_{rR}(x^0)) \tag{7.2}$$

as is seen by a simple calculation. We are interested in what sort of functions are subsequential limits as $r \downarrow 0$ and $r \uparrow \infty$ (in which case we need $U = \mathbb{R}^n$).

Proposition 7.1.

(a) Let u be ∞-harmonic on $B_1(x^0)$. Then the set of functions $\{v_r : 0 < r < 1/(2R)\}$ is precompact in $C(B_R(0))$ for each $R > 0$ and if $r_j \downarrow 0$ and $v_{r_j} \to v$ locally uniformly on $\mathbb{R}^n$, then $v(x) = \langle p, x \rangle$ for some p satisfying $|p| = S^+(x^0)$.

(b) Let u be ∞-harmonic on $\mathbb{R}^n$ and $\mathrm{Lip}(u, \mathbb{R}^n) < \infty$. Then the set of functions $\{v_r : r > 0\}$ is precompact in $C(\mathbb{R}^n)$ and if $R_j \uparrow \infty$ and $v_{R_j} \to v$ locally uniformly on $\mathbb{R}^n$, then $v(x) = \langle p, x \rangle$ for some p satisfying $|p| = \mathrm{Lip}(u, \mathbb{R}^n)$.

Proof. We prove (b), which is new, and relegate (a), which is known, to Exercise 7.1 below. Let the assumptions of (b) hold. First we notice that

$$\mathrm{Lip}(u, \mathbb{R}^n) = \lim_{R \to \infty} \frac{\max_{\overline{B}_R} u}{R}. \tag{7.3}$$

Indeed, for every $x \in \mathbb{R}^n$,

$$L(u, x) = S^+(x) \leq \max_{w \in \overline{B}_R(x)} \frac{u(w) - u(x)}{R}$$

$$\leq \max_{w \in B_{R+|x|}(0)} \frac{u(w)}{R} - \frac{u(x)}{R} \to \lim_{R \to \infty} \frac{\max_{\overline{B}_R} u}{R} \tag{7.4}$$

as $R \to \infty$. This proves (7.3) with "$\leq$" in place of the equal sign. Letting x_R be a maximum point of u relative to $\partial B_R(0)$, we also have

$$\frac{\max_{\overline{B}_R} u}{R} = \frac{u(x_R) - u(0)}{R} + \frac{u(0)}{R} \leq \mathrm{Lip}(u, \mathbb{R}^n) + \frac{u(0)}{R} \to \mathrm{Lip}(u, \mathbb{R}^n),$$

which provides the other side of (7.3).

Now let $r = R_j$ where $R_j \uparrow \infty$. We set $x^0 = 0$ and now denote $v_j := v_{R_j}$. Clearly $\mathrm{Lip}(v_j, \mathbb{R}^n) \leq \mathrm{Lip}(u, \mathbb{R}^n)$. Passing to a subsequence for which $v_j \to v$ locally uniformly, we claim that v is linear. Clearly $v(0) = 0$. As u is ∞-subharmonic, for $R > 0$ there is a unit vector x_R^j such that

$$v_j(Rx_R^j) = \max_{B_R(0)} v_j = \frac{\max_{\overline{B}_{R_j R}} u - u(0)}{R_j}.$$

Passing to a subsequence along which $x_R^j \to x_R^+ \in \partial B_1(0)$ and taking the limit $j \to \infty$ above, we have, via (7.3),

$$v(Rx_R^+) = R\mathrm{Lip}(u, \mathbb{R}^n) \geq R\mathrm{Lip}(v, \mathbb{R}^n).$$

As u is also ∞-superharmonic, there is also a unit vector x_R^-, such that

$$v(Rx_R^-) = -R\mathrm{Lip}(u, \mathbb{R}^n) \leq -R\mathrm{Lip}(v, \mathbb{R}^n).$$

It follows that

$$2R\mathrm{Lip}(v, \mathbb{R}^n) \leq 2R\mathrm{Lip}(u, \mathbb{R}^n) = v(x_R^+) - v(x_R^-) \leq R\mathrm{Lip}(v, \mathbb{R}^n)|x_R^+ - x_R^-|.$$

If $\mathrm{Lip}(u, \mathbb{R}^n) = 0$, then $v \equiv 0$ and we are done. If $\mathrm{Lip}(u, \mathbb{R}^n) > 0$, then we deduce that $|x_R^+ - x_R^-| = 2$, which implies that $x_R^+ = -x_R^-$, as well as $\mathrm{Lip}(v, \mathbb{R}^n) = \mathrm{Lip}(u, \mathbb{R}^n)$. In particular, the points x_R^+, x_R^- are unique. By (2.10), v is linear on each segment $[Rx_R^-, Rx_R^+]$, and this implies, together with the uniqueness, that $x_R^-, x_R^+ = \omega$ is a unit vector independent of R. Altogether we have

$$v(t\omega) = t\mathrm{Lip}(v, \mathbb{R}^n) = t\mathrm{Lip}(u, \mathbb{R}^n).$$

The assertion (b) now follows from Section 7.2 below.

Exercise 7.1. Prove, by similar arguments, (a) of the proposition.

Definition 7.1. *In the case (a) of Proposition 7.1, if $r_j \downarrow 0$ and $v_{r_j} \to \langle p, x \rangle$ on $\mathbb{R}^n$, we call p a derivate of u at x^0. Similarly, in case (b), one defines "derivates of u" at ∞.*

Exercise 7.2. Show, in case (a) of Proposition 7.1, that if the set of derivates of u at x^0 consists of a single point, then u is differentiable at x^0. Conclude that if $S^+(x^0) > 0$, then u is differentiable at x^0 if and only if

$$1 = |\omega_r| \text{ and } u(x^0 + r\omega_r) = \max_{\overline{B}_r(x^0)} u,$$

then $\lim_{r \downarrow 0} \omega_r$ exists.

Exercise 7.3. Use (a) of Proposition 7.1 to show that

$$\lim_{r \downarrow 0} \min_{|p|=S^+(x^0)} \max_{|x|\le r} \left| \frac{u(x_0 + x) - u(x_0) - \langle p, x \rangle}{r} \right| = 0.$$

7.2 Implications of Tight on a Line Segment

Please forgive us the added generality we are about to explain. The costs aren't large, and we wanted to share this little proof. It is a straightforward generalization and extension of a proof learned from [3] in the Euclidean case. However, the reader may just assume that $|\cdot|^* = |\cdot|$ and $F(x) = x/|x|$ if $x \neq 0$ below to restrict to the Euclidean case.

In this section we use the notation $|x|$, as usual, to denote the norm of $x \in \mathbb{R}^n$. However, the norm *need not be the Euclidean norm*. However, we do assume that $x \to |x|$ is differentiable at any $x \neq 0$. This is equivalent to the dual norm, denoted by $|\cdot|^*$, being strictly convex, where

$$|x|^* := \max\{\langle x, y \rangle : |y| \le 1\} \tag{7.5}$$

and $\langle x, y \rangle$ denotes the usual inner-product of $x, y \in \mathbb{R}^n$. The duality map F (from $(\mathbb{R}^n, |\cdot|)$ to $(\mathbb{R}^n, |\cdot|^*)$) is defined by

$$\begin{cases} |F(x)|^* = 1 \quad \text{and} \quad |x| = \langle x, F(x) \rangle & \text{if} \quad x \neq 0 \\ F(0) = \{y : |y|^* \le 1\}. \end{cases} \tag{7.6}$$

When $x \neq 0$ there is only one vector with the properties assigned to $F(x)$ because of the strict convexity of $|\cdot|^*$; moreover, if $x \neq 0$,

$$|x + y| = |x| + \langle y, F(x) \rangle + o(y) \quad \text{as} \quad y \to 0. \tag{7.7}$$

This follows from the relation

$$\frac{d}{dt}|x + ty| = \langle y, F(x + ty) \rangle \tag{7.8}$$

when $x + ty \neq 0$ and continuity of F away from 0.

Lemma 7.1. *Let $U \subset \mathbb{R}^n$ be open and $u : U \to \mathbb{R}$ satisfy*

$$|u(x) - u(y)| \le |x - y| \quad \text{for} \quad x, y \in U. \tag{7.9}$$

Let $\mathcal{I}$ be an open interval containing 0, $x^0 \in U$, $p \in \mathbb{R}^n$, $|p| = 1$ and $x^0 + tp \in U$ for $t \in \mathcal{I}$. Let

$$u(x^0 + tp) = u(x^0) + t \quad \text{for} \quad t \in \mathcal{I}. \tag{7.10}$$

Then:

(a) u is differentiable at x^0 and $Du(x^0) = F(p)$.

(b) If also $\mathcal{I} = \mathbb{R}$, then u is the restriction to U of the affine function

$$v(x^0 + x) = u(x^0) + \langle x, F(p) \rangle$$

on $\mathbb{R}^n$.

Proof. We may assume that $x^0 = 0$ and $u(x^0) = 0$ without loss of generality. Let P be "projection" along p;

$$Px = \langle x, p^* \rangle\, p \quad \text{where} \quad p^* = F(p).$$

For small $x \in U$ and small $r \in \mathbb{R}$, (7.9) and (7.10) yield

(i) $\langle x, p^* \rangle + r - |(x - Px) - rp| = u(Px + rp) - |(x - Px) - rp| \le u(x),$
(ii) $u(x) \le u(Px + rp) + |(x - Px) - rp| = \langle x, p^* \rangle + r + |(x - Px) - rp|.$

$$\tag{7.11}$$

These relations are valid whenever $\langle x, p^* \rangle \pm r \in \mathcal{I}$ and $x \in U$. Rearranging (7.11) yields

$$r - |(x - Px) + rp| \le u(x) - \langle x, p^* \rangle \le r + |(x - Px) + rp|. \tag{7.12}$$

We seek to estimate the extreme expressions above. To make the left-most extreme small, we take $r > 0$. Then, by differentiability of $|\cdot|$,

$$|(x - Px) - rp| - r = r\left(\left| p - \frac{x - Px}{r} \right| - 1 \right)$$

$$= r\left(|p| - \frac{1}{r}\langle x - Px, p^* \rangle - 1 \right) + ro\left(\frac{|x - Px|}{r} \right) \tag{7.13}$$

$$= ro\left(\frac{|x - Px|}{r} \right).$$

Here we used $|p| = 1$ and $\langle x - Px, p^* \rangle = \langle x, p^* \rangle - \langle x, p^* \rangle \langle p, p^* \rangle = 0$. We use (7.13) two ways. First, if r is bounded and bounded away from 0, we have

$$ro\left(\frac{|x - Px|}{r} \right) = o(|x - Px|) = o(|x|) \quad \text{as} \quad x \to 0.$$

To make the right-most extreme of (7.12) small, we take $r < 0$ and proceed similarly. Returning to (7.12), this verifies the differentiability of u at 0 and $Du(0) = p$. On the other hand, if we may send $|r| \to \infty$, the extremes of (7.13) tend to 0. Returning to (7.12), this verifies $u(x) = \langle x, p^* \rangle$ in U.

Exercise 7.4. Let $|\cdot|$ be the Euclidean norm. Let K be a closed subset of $\mathbb{R}^n$. Suppose $z \in \mathbb{R}^n \setminus K$ and

$$y \in K \text{ and } |w - y| \le |w - z| \text{ for } z \in K.$$

Show that $x \mapsto \text{dist}\,(x, K)$ is differentiable at each point of $[y, w)$ and compute its derivative. Give an example where differentiability fails at w.

Exercise 7.5. Let $U = \mathbb{R}^n \setminus \partial U$. Let $b \in C(\partial U)$ and $L := \mathrm{Lip}(b, \partial U) < \infty$. Show that if $x \in U$ and $\mathcal{MW}_*(b)(x) = \mathcal{MW}^*(b)(x)$, then both $\mathcal{MW}_*(b)$ and $\mathcal{MW}_*(b)$ are differentiable at x. Show that $\mathcal{MW}_*(b) = \mathcal{MW}^*(b)$ if and only if they are both in $C^1(U)$ and satisfy the eikonal equation $|Du| = L$ in U. Hint: For the first part, show that x lies in the line segment $[y, z]$ when $y, z \in \partial U$ and $b(y) - L|y - x| = b(z) + L|z - x|$.

8 An Impressionistic History Lesson

The style of this section is quite informal; we seek to convey the flow of things, hopefully with enough clarity, but without distracting precision. It is assumed that the reader has read the introduction, but not the main text of this article. We do include some pointers, often parenthetical, to appropriate parts of the main text.

8.1 The Beginning and Gunnar Aronosson

It all began with Gunnar Aronsson's paper 1967 paper [3]. The functional Lip is primary in this paper, but two others are mentioned, including $\mathcal{F}_\infty$. Aronsson observed that $\mathrm{Lip} = \mathcal{F}_\infty$ if U is convex, while this is not generally the case if U is not convex.

The problem of minimizing $\mathcal{F} = \mathrm{Lip}$ subject to a Dirichlet conditions was known to have a largest and a smallest solution, given by explicit formulas ($\mathcal{MW}_*, \mathcal{MW}^*$ of (2.2)), via the works of McShane and Whitney [44], [51]. Aronsson derived, among other things, interesting information about the set on which these two functions coincide (Exercise 7.5)) and the derivatives of any solution on this "contact set". In particular, he established that minimizers for Lip are unique iff there is a function $u \in C^1(U) \cap C(\overline{U})$ which satisfies

$$|Du| \equiv \mathrm{Lip}(b, \partial U) \text{ in } U, \ u = b \text{ on } \partial U,$$

which is then the one and only solution. This is a very special circumstance. Moreover, in general, the McShane-Whitney extensions have a variety of unpleasant properties (Exercise 2.1). The following question naturally arose: is it possible to find a canonical Lipschitz constant extension of b into U that would enjoy comparison and stability properties? Furthermore, could this special extension be unique once the boundary data is fixed? The point of view was that the problem was an "extension" problem - the problem of extending the boundary data b into U without increasing the Lipschitz constant, hopefully in a manner which had these other good properties. Aronsson's - eventually successful - proposal in this regard was to introduce the class of absolutely minimizing functions for Lip, which generalized notions already appearing in works of his in one dimension ([1], [2]). Aronsson further gave the outlines of an existence proof not so different from the one sketched in Section 5, but using the McShane-Whitney extensions rather than cones. This required the boundary data to be Lipschitz continuous (in contrast with Theorem 5.1). He could not, however, prove the uniqueness or stability (Theorem 5.2).

Thinking in terms of $\mathcal{F}_\infty = \mathrm{Lip}$ on convex sets, Aronsson was led to the now famous pde:

$$\Delta_\infty u = \sum_{i,j=1}^{n} u_{x_i} u_{x_j} u_{x_i x_j} = 0.$$

He discovered this by heuristic reasoning: first, by standard reasoning, if $1 < p$, then

u minimizes $\mathcal{F}_p(u, U) := \||Du|\|_{L^p(U)}\, dx$ among functions

satisfying $u = b$ on ∂U iff $u = b$ on ∂U and $\dfrac{1}{p-2}|Du|^2 \Delta u + \Delta_\infty u = 0$,

provided that $\mathcal{F}_p(u, U) < \infty$.

Letting $p \to \infty$ yields $\Delta_\infty u = 0$. Moreover, $\mathcal{F}_p(u, U) \to \mathcal{F}_\infty(u, U)$ if $|Du| \in L^\infty(U)$.

Aronsson further observed that for $u \in C^2$, $\Delta_\infty u = 0$ amounts to the constancy of $|Du|$ on the lines of the gradient flow (Section 2.4). He went on to prove that if $u \in C^2(U)$, then u is absolutely minimizing for Lip in U iff $\Delta_\infty u = 0$ in U (Section 2.4).

With the technology of the times, this is about all anyone could have proved. The gaps between $\Delta_\infty u = 0$ being the "Euler equation" if $u \in C^2$ and his existence proof, which produced a function only known to be Lipschitz continuous, could not be closed at that time. In particular, Aronsson already knew that classical solutions of the eikonal equation $|Du| = $ constant, which might not be C^2, are absolutely minimizing. However, he offered no satisfactory way to interpret them as solutions of $\Delta_\infty u = 0$. Moreover, the question of uniqueness of the function whose existence Aronsson proved would be unsettled for 26 years!

Aronsson himself made the gap more evident in the paper [4] in which he produced examples of U, b for which he could show that the problem had no C^2 solution. This work also contained a penetrating analysis of classical solutions of the pde. However, all of these results are false in the generality of *viscosity solutions* of the equation (see below), which appear as the perfecting instrument of the theory.

The best known explicit irregular absolutely minimizing function - outside of the relatively regular solutions of eikonal equations - was exhibited again by Aronsson, who showed in the 1984 paper [6] that $u(x, y) = x^{4/3} - y^{4/3}$ is absolutely minimizing in R^2 for Lip and $\mathcal{F}_\infty$ (Exercises 2.6, 2.7).

Most of the interesting results for classical solutions in 2 dimensions proved by Aronsson are falsified by this example. These include: $|Du|$ is constant on trajectories of the gradient flow, global absolutely minimizing functions are linear, and Du cannot vanish unless u is locally constant. A rich supply of other solutions was provided as well.

8.2 Enter Viscosity Solutions and R. Jensen

Let $u \in \text{USC}(U)$ (the upper semicontinuous functions on U). Then u is a viscosity subsolution of $\Delta_\infty u = 0$ (equivalently, a viscosity solution of $\Delta_\infty u \geq 0$) in U if: whenever $\varphi \in C^2(U)$ and $u - \varphi$ has a local maximum at $\hat{x} \in U$, then $\Delta_\infty \varphi(\hat{x}) \geq 0$.

Let $u \in \text{LSC}(U)$. Then u is a viscosity supersolution of $\Delta_\infty u = 0$ (equivalently, a viscosity solution of $\Delta_\infty u \leq 0$) in U if whenever $\varphi \in C^2(U)$ and $u - \varphi$ has a local minimum at $\hat{x} \in U$, then $\Delta_\infty \varphi(\hat{x}) \leq 0$.

The impetus for this definition arises from the standard maximum principle argument at a point $\hat{x}$ where $u - \varphi$ has a local maximum. Let $D^2 u = \left(u_{x_i, x_j}\right)$ be the Hessian matrix of the second order partial derivatives of u. Then

$$\Delta_\infty u = \langle D^2 u Du, Du \rangle \geq 0, \; Du(\hat{x}) = D\varphi(\hat{x}) \text{ and } D^2 u(\hat{x}) \leq D^2 \varphi(\hat{x})$$
$$\implies \langle D^2 \varphi(\hat{x}) D\varphi(\hat{x}), D\varphi(\hat{x}) \rangle \geq 0.$$

This puts the derivatives on the "test function" φ via the maximum principle, a device used by L. C. Evans in 1980 in [32].

The theory of viscosity solutions of much more general equations, born in first order case the 1980's, and to which Jensen made major contributions, contained strong results of the form:

Comparison Theorem: *Let $u, -v \in \text{USC}(\overline{U})$,*
$$G(x, u, Du, D^2 u) \geq 0 \text{ and } G(x, v, Dv, D^2 v) \leq 0 \text{ in } U$$
in the viscosity sense, and $u \leq v$ on ∂U. Then $u \leq v$ in U.

The developers of these results in the second order case were Jensen, Ishii, Lions, Souganidis, $\cdots$. They are summarized in [31], which contains a detailed history. In using [31], note that we are thinking here of G being an increasing function of the Hessian, $D^2 u$, not a decreasing function. Hence solutions of $G \geq 0$ are subsolutions.

Of course, there are structure conditions needed on G in order that the Comparison Theorem hold in addition to standard maximum principle enabling assumptions, and these typically imply that the solutions of $G \geq 0$ perturb to a solution of $G > 0$ or that some change of variables such as $u = g(w)$ produces a new problem with this property. This had not been not accomplished in any simple way for Δ_∞, except at points where "$Du \neq 0$" until the method explained in Section 5, which gives an approximation procedure for subsolutions which produces subsolutions with nonvanishing gradients, was developed. This is coupled with a change of variable such as $u = w - \frac{\lambda}{2} w^2$, under which $\Delta_\infty u \geq 0$ implies $\Delta_\infty w \geq \frac{\lambda}{1 - \lambda w} |Dw|^4$.

Moreover, Hitoshi Ishii introduced the Perron method in the theory of viscosity solutions, which provides "existence via uniqueness," that is, roughly speaking,

Existence Theorem: *When the Comparison Theorem is true and there exist u, v satisfying its assumptions, continuous on $\overline{U}$ and satisfying*
$$u = b = v \text{ on } \partial U$$
then there exists a (unique) viscosity solution $u \in C(\overline{U})$ of
$$0 \leq G(x, u, Du, D^2 u) \leq 0 \text{ in } U \text{ and } u = b \text{ on } \partial U$$

See Section 4 of [31].

We wrote $0 \leq G \leq 0$ above to highlight that a viscosity solution of $G = 0$ is exactly a function which is a viscosity solution of both $G \leq 0$ AND $G \geq 0$; there is no other notion of $G = 0$ in the viscosity sense.

NOTE: hereafter all references to "solutions", "subsolutions", $\Delta_\infty u \leq 0$, etc, are meant in the viscosity sense!

As mentioned, in 1993 R. Jensen proved, in [36], that

(J1) Absolute minimizers u for $\mathcal{F}_\infty$ are characterized by $\Delta_\infty u = 0$.

(J2) The comparison theorem holds for $G = \Delta_\infty$.

Jensen's proof of the comparison theorem was remarkable. In order to deal with the difficulties associated with points where $Du = 0$, he used approximations via the "obstacle problems"

$$\max \left\{ \varepsilon - |Du^+|, \Delta_\infty u^+ \right\} = 0, \ \min \left\{ |Du^-| - \varepsilon, \Delta_\infty u^- \right\} = 0!$$

These he "solved" by approximation with modifications of $\mathcal{F}_p$ and then letting $p \to \infty$, although they are amenable to the general theory discussed above. It is easy enough to show that

$$u^- \leq u \leq u^+$$

when $\Delta_\infty u = 0$ and $u = u^+ = u^- = b$ on ∂U. Comparison then followed from an estimate, involving Sobolev inequalities, which established $u^+ - u^- \leq \kappa(\varepsilon)$ where $\kappa(0+) = 0$.

The first assertion of (J1) was proved directly, via a modification of Aronsson's original proof, while the "conversely" was a consequence of existence and uniqueness. The relation between "absolutely minimizing" relative to $\mathcal{F}_\infty$ and relative to Lip had become even more murky. Jensen referred as well to another "$\mathcal{F}$", as had Aronsson earlier, namely the Lipschitz constant relative to the "interior distance" between points:

$$\text{dist}_U(x, y) = \text{infimum of the lengths of paths in } U \text{ joining } x \text{ and } y$$

and the ordinary Lipschitz constant did not play a role in his work.

Thus, after 26 years, the existence of absolutely minimizing functions assuming given boundary values was known (Aronsson and Jensen), and, at last, the uniqueness (Jensen).

Jensen's work generated considerable interest in the theory. Among other contributions was an lovely new uniqueness proof by G. Barles and J. Busca in [9]. Roughly speaking, this proof couples some penetrating observations to the standard machinery of viscosity solutions to reach the same conclusions as Jensen, but without obstacle problems or integral estimates. This was the state of the art until the method of Section 5 was deployed.

After existence and uniqueness, one wants to know about regularity.

8.3 Regularity

Aronsson's example, $u(x, y) = x^{4/3} - y^{4/3}$, sets limits on what might be true. The first derivatives of u are Hölder continuous with exponent $1/3$; second derivatives do not exist on the lines $x = 0$ and $y = 0$. It is still not known whether or not every ∞-harmonic function has this regularity.

Modulus of Continuity

The first issue is the question of a modulus of continuity for absolutely minimizing functions. In Aronsson's framework, he dealt only with locally Lipschitz continuous functions. In our Lemma 4.1 we establish the local Lipschitz continuity of USC ∞-harmonic functions via comparison with cones. Jensen also gave similar arguments to establish related results. See also [19]. As Lipschitz continuous functions are differentiable almost everywhere, so are ∞-harmonic functions.

Harnack and Liouville

Aronsson's original "derivation" of $\Delta_\infty u = 0$ as the "Euler equation" corresponding to the property of being absolutely minimizing and Jensen's existence and uniqueness proofs closely linked letting $p \to \infty$ in the problem

$$\Delta_p u_p := |Du|^{p-4} \left(|Du|^2 \Delta u + (p-2)\Delta_\infty u \right) = 0 \text{ in } U$$

and $u_p = b$ on ∂U with our problem

$$\Delta_\infty u = 0 \text{ in } U \text{ and } u = b \text{ on } \partial U.$$

With this connection in mind and using estimates learned from the theory of Δ_p, P. Lindqvist and J. Manfredi [42] proved that if $u \geq 0$ is a *variational* solution of $\Delta_\infty u \leq 0$ (i.e, a limit of solutions of $\Delta_p u_p \leq 0$), then one has the *Harnack Inequality*

$$u(x) \leq e^{\frac{|x-y|}{R-r}} u(y) \text{ for } x, y \in B_r(x_0) \subset B_R(x_0) \subset U.$$

They derived this from the elegant *Gradient Estimate*

$$|Du(x)| \leq \frac{1}{\text{dist}\,(x, \partial U)} (u(x) - \inf_U u),$$

valid at points where $Du(x)$ exists. (Our equation (4.20) is more precise and the proof is considerably more elementary.) The appropriate Harnack inequality is closely related to regularity issues for classes of elliptic and parabolic equations, which is one of the reasons to be interested in it. However, so far, it has not played a similar role in the theory of Δ_∞. In particular, note that the gradient estimate implied the Harnack inequality in the reasoning of [42]; estimating the gradient came first.

The same authors extended this result, a generalization of an earlier result of Evans for smooth functions, to all ∞-superharmonic functions, ie, solutions of $\Delta_\infty u \leq 0$, in [43], by showing that ALL ∞-superharmonic functions are variational. This perfected the relationship between $\Delta_p, \mathcal{F}_p$ and $\Delta_\infty, \mathcal{F}_\infty$. By the way, the original observation that if solutions of $\Delta_p u_p = 0$ have a limit $u_p \to u$ as $p \to \infty$, then $\Delta_\infty u = 0$ is due to Bhattacharya, Di Benedetto and Manfredi, [24] (1989). This sort of observation is a routine matter in the viscosity solution theory; there is much else in that paper. One point of concern is the relationship between the notions of viscosity solutions and solutions in the sense of distributions for the p-Laplace equation. See Juutinen, Lindqvist and Manfredi, [39] and Ishii, [35].

Lindqvist and Manfredi also showed that

(LM1) If $u(x) - v(x) \geq \min_{\partial V}(u - v)$ when $\Delta_\infty v = 0$, and $x \in V \subset\subset U$,

then $\Delta_\infty u \leq 0$.

(LM2) If $\Delta_\infty u = 0$ in $\mathbb{R}^n$ and u is bounded below, then u is constant.

Comparison with Cones, Full Born

Subsequently, Crandall, Evans and Gariepy [29] showed that it suffices to take functions of the form

$$v(x) = a|x - z|, \quad \text{also known as a "cone function",}$$

where $z \notin V$, in (LM1), and introduced the terminology "comparison with cones" (Definition 2.2 and Theorem 2.1).

The assumptions of (LM1) with v a cone function as above is called comparison with cones from below; the corresponding relation for $\Delta_\infty u \geq 0$ is called comparison with cones from above. When u enjoys both of these, it enjoys comparison with cones. All the information contained in $\Delta_\infty u \leq 0, \Delta_\infty u \leq 0, \Delta_\infty u = 0$ is contained in the corresponding comparison with cones property.

Also proved in [29], with a 2-cone argument, was the generalization

$$\Delta_\infty u \leq 0 \text{ and } u(x) \geq a + \langle p, x \rangle \text{ in } \mathbb{R}^n \implies u(x) = u(0) + \langle p, x \rangle.$$

of (LM2). With this generality, it follows that if $\Delta_\infty u \leq 0$, $\varphi \in C^1$ and $\hat{x}$ is a local minimum of $u - \varphi$, then u is differentiable at $\hat{x}$. At last: a result asserting the existence of a derivative at a particular point.

More importantly, it had become clear that approximation by Δ_p is probably not the most efficient path to deriving properties of ∞-sub and super harmonic and ∞-harmonic functions. Of course, comparison with cones had already been used by Jensen to derive Lipschitz continuity, and was used contemporaneously with [29] by Bhattacharya [19], etc., but it was now understood that this approach made use of all the available information.

Recall that in Jensen's organization, he showed that if $\Delta_\infty u = 0$ fails, then u is not absolutely minimizing for $\mathcal{F}_\infty$; equivalently, if u is absolutely minimizing, then $\Delta_\infty u = 0$. Then he used existence/uniqueness in order to establish the converse: if $\Delta_\infty u = 0$, then u is absolutely minimizing. In [29] it was proved directly - without reference to existence or uniqueness - that comparison with cones, and hence $\Delta_\infty u = 0$, implies that u is absolutely minimizing for $\mathcal{F}_\infty$ (Proposition 6.1).

Blowups are Linear

The next piece of evidence in the regularity mystery was provided by Crandall and Evans, [30]. Using tools from [29] and some new arguments, all rather simple.

They proved that if $\Delta_\infty u = 0$ near x_0 and

$$r_j \downarrow 0, \; v_j(x) = \frac{u(x_0 + r_j x) - u(x_0)}{r_j}, \; v(x) = \lim_{j \to \infty} v_j(x)$$

then $v(x) = \langle p, x \rangle$ for some $p \in \mathbb{R}^n$ (Section 7).

Note that since u is Lipschitz in each ball $B_R(x_0)$ in its domain, each v_j is Lipschitz with the same constant in $B_{R/r_j}(0)$. Thus any sequence $r_j \downarrow 0$ has a subsequence along which the v_j converge locally uniformly in $\mathbb{R}^n$. In consequence, if x_0 is a point for which

$$|z_r - x_0| = r, u(z_r) = \max_{\overline{B}_r(x_0)} u \implies \lim_{r \downarrow 0} \frac{z_r - x_0}{|z_r - x_0|} \text{ exists,}$$

then u is differentiable at x_0. However, no one has been able to show that the maximum points have a limiting direction, except as a consequence of Ovidiu Savin's results in [49], in the case $n = 2$.

Savin's Theorem

Savin [49] showed that ∞-harmonic functions are C^1 if $n = 2$. Savin does not work on the "directions" of $z_r - x_0$ mentioned above directly. He does start from a reformulation of the "$v(x) = \langle p, x \rangle$" result above (Exercise (7.3)). It does not appear that Savin's arguments contain any clear clues about the case $n > 2$, as he uses the topology of $\mathbb{R}^2$ very strongly, and the question of whether ∞-harmonic functions are necessarily C^1 in general remains the most prominent open problem in the area. Moreover, while Savin provides a modulus of continuity for Du when u is ∞-harmonic, this modulus is not explicit. It would be quite interesting to have more information about it. Is it Hölder 1/3, as for $x^{4/3} - y^{4/3}$? Probably not. Savin shows, in consequence of his other results, that if u is ∞-harmonic on R^2 and globally Lipschitz continuous, then u is affine. The corresponding question in the case $n > 2$ is also open, and it is certainly related to the C^1 question.

9 Generalizations, Variations, Recent Developments and Games

First of all, there is by now a substantial literature concerning optimization problems with supremum type functionals. Much of this theory was developed by N. Barron and R. Jensen in collaboration with various coauthors. We refer the reader to the review article by Barron [10] for an overview up to the time of its writing, and to Jensen, Barron and Wang [11], [12] for more recent advances. In particular, [12] is concerned with vector-valued functions u in the set up we explain below for scalar functions u. However, in the vector case, existence of minimizers and not absolute minimizers is the focus (unless $n = 1$).

We take the following point of view in giving selected references here. If one goes to MathSciNet, for example, and brings up the review of Jensen's paper [36], there will be over 30 reference citations (which is a lot). This will reveal papers with titles involving homogenization, Γ-limits, eigenvalue problems, free boundary problems, and so on. None of these topics are mentioned in this work of limited aims. Likewise, one can bring up a list of the papers coauthored by Barron and/or Jensen, etc., or any of the authors which popped up as referencing Jensen [36], and then search the web to find the web sites of authors of article that interest you, it is a new world. Perhaps it is worth mentioning that the Institute for Scientific Information's *Web of Science* generally provides a more complete cited reference search (Jensen's article gets over 50 citations on the Web of Science).

So we are selective, sticking to variants of the main thrust of this article.

9.1 What is Δ_∞ for $H(x, u, Du)$?

A very natural generalization of the theory of the preceding sections arises by replacing the functional $\mathcal{F}_\infty$ by a more general functional

$$\mathcal{F}_\infty^H(u, U) := \|H(x, u, Du(x))\|_{L^\infty(U)} \tag{9.1}$$

for suitable functions H. We write the generic arguments of H as $H(x, r, p)$. H should be reasonable, and for us $r \in \mathbb{R}$ is real, corresponding to $u : U \to \mathbb{R}$. The case discussed in these notes is $H(x, r, p) = |p|$, but we could as well put $H(x, r, p) = |p|^2$, which has the virtue that H is now smooth, along with being convex and quite coercive. It turns out that for part of the theory, it is not convexity of H which is primary, but instead "quasi-convexity," which means that each sublevel set of H is convex:

$$\{p : H(x,r,p) \leq \lambda\} \text{ is convex for each } x \in \overline{U}, r, \lambda \in \mathbb{R}. \qquad (9.2)$$

We are going to suppress more technical assumptions on H, such as the necessary regularity, coercivity, and so on, needed to make statements precise in most of this discussion. The reader should go to the references given for this, if it is omitted.

The operator corresponding to Δ_∞ in this generality is

$$\mathcal{A}(x,r,p,X) = \langle H_x(x,r,p) + H_r(x,r,p)p + XH_p(x,r,p), H_p(x,r,p)\rangle. \qquad (9.3)$$

By name, we call this the "Aronsson operator" associated with H. It is defined on arguments (x,r,p,X), where X is a symmetric $n \times n$ real matrix. The notations H_x, H_p stand for the gradients of H in the x and p variables, while H_r is $\partial H/\partial r$. The Aronsson equation is $\mathcal{A}[u] := \mathcal{A}(x, u, Du, D^2u) = 0$. In this form, it is more easily remembered as

$$\mathcal{A}[u] = \langle H_p(x, u(x), Du(x)), D_x(H(x, u(x), Du(x)))\rangle = 0.$$

Observe that if $H = (1/2)|p|^2$, then $\mathcal{A}[u] = \Delta_\infty u$, while if $H = |p|$ we would have instead

$$\mathcal{A}(x,r,p,X) = \langle X\hat{p}, \hat{p}\rangle \text{ where } \hat{p} = \frac{p}{|p|}. \qquad (9.4)$$

There is a viscosity interpretation of equations with singularities such as (9.4), and at $p = 0$ this interpretation just leads to our relations (2.24), (2.26). It was shown by Barron, Jensen and Wang [11] that if u is absolutely minimizing for $\mathcal{F}_\infty^H$, then $\mathcal{A}[u] = 0$ in the viscosity sense. The technical conditions under which these authors established this are more severe than those given in [27], corresponding to the more transparent proof given in this paper. It remains an interesting question if one assumption common to [11], [27] can be removed, namely, is it sufficient to have $H \in C^1$ (rather than C^2)?

It remained a question as to whether or not $\mathcal{A}[u] = 0$ implied that u is absolutely minimizing for $\mathcal{F}_\infty^H$. Y. Yu [53] proved several things in this direction. First, if $H = H(x,p)$ is convex in p and sufficiently coercive, the answer is yes. Secondly, he provided an example to show that the answer is no in general if H is merely quasi-convex, but otherwise nice enough. He takes $n = 1$, $H(x,p) = (p^2 - 2p)^3 + V(x)$ and designs V to create the counterexample. Likewise, Yu showed that in the case $H = H(r,p)$, the Aronsson equation does not guarantee absolutely minimizing. Subsequently, Gariepy, Wang and Yu [34] showed that if $H = H(p)$, that is, H does not depend on x, r, and is merely quasi-convex, then, indeed, $\mathcal{A}[u] = 0$ implies absolutely minimizing.

Moreover, Yu also showed that there is no uniqueness theorem in the generality of $\mathcal{F}_\infty^H$. This is not an issue of smoothness of H. Yu gives the simple example of $n = 1$, $U = (0, 2\pi)$, $H(x, p) = |p|^2 + \sin^2(x)$ and notes that both $u \equiv 0$ and $u = \sin(x)$ solve

$$\mathcal{A}[u] = 2u_x(2\cos(x)\sin(x) + 2u_xu_{xx}) = 0, \quad u(0) = u(2\pi) = 0.$$

9.2 Generalizing Comparison with Cones

Are there cones in greater generality than in the archetypal case $H = |p|$ and $|\cdot|$ is any norm on $\mathbb{R}^n$? Note that in this case, if $a \geq 0$, we have the cone functions

$$a|x - z|^* = \sup_{H(p)=a} \langle x - z, p \rangle,$$

where $|\cdot|^*$ is the dual norm. That is, with this H, absolutely minimizing for $\mathcal{F}_\infty^H$ couples with comparison with cones defined via $|\cdot|^*$ and absolutely minimizing for Lip where Lipschitz constants are computed with respect to this dual norm. This is the generality of [8]. In this case, u is absolutely minimizing for $\mathcal{F}_\infty^H$ if and only if u and $-u$ enjoy comparison with these cone functions from above, per [8], and these properties are equivalent to u being a solution of the Aronsson equation if $|\cdot|^*$ and $|\cdot|$ are C^1 off of the origin, and in other cases as well.

In the general quasi-convex case in which $H(p)$ is C^2 and coercive, Gariepy, Wang and Yu show that the same formula yields a class of functions C_H^a $(x-z)$ which has the same properties. One loses some norm-like properties and the generality of the situation requires a more delicate analysis to show that comparison with these generalized cones implies that the Aronsson equation is satisfied. Other arguments given herein do not easily generalize to this case.

The case in which $H(x, p)$ is convex in p, along with other technical assumptions, Yu [53] uses comparison type arguments to show similar results.

The general case $H(x, p)$ is treated as well, in full quasi-convex generality and with minimum regularity on H, in Champion and De Pascale, [25]. These authors supply a "comparison with distance functions" equivalence for the property of being absolutely minimizing with respect to $\mathcal{F}_\infty^H$. This is a bit too complex to explain here (while quite readable in [25]), so we content ourselves by noting that this property is defined by Mc Shane - Whitney type operators relative to distance within a given set, coupled with appropriate structures related to the definition of generalized cones just above. There are other interesting things in this paper.

One key difference between works in this direction and the simpler case treated in this paper, is that they do not use any evident analogue of the interplay between the functionals Lip and $\mathcal{F}_\infty$, which we used to help our organization.

9.3 The Metric Case

Let us take this case to include abstract metric spaces as well as metrics arising from differential geometric considerations. In the abstract case, we mention primarily Champion and De Pascale [25], Section 5, as this seems to be the state of the art in this direction. Here the Lipschitz functional is primary, and the authors show that the associated absolutely minimizing property is equivalent to a straightforward comparison with distance functions property. In particular, it makes clear that the fact that the distance functions do not themselves satisfy a full comparison with distance function property noted in [8] is not an impediment to the full characterization, properly put. Existence had previously been treated in Jutinen [38] and Mil'man [45], [46].

For papers which treat various geometrical structures, we mention the the work Bieske [15], which was the first, as well as [16], [17], [18] and Wang [50]. The paper of Wang, currently available on his website, contains a very nice introduction to which we refer for a further overview.

9.4 Playing Games

It was recently discovered by Peres, Schramm, Sheffield and Wilson [48] that the value function for a random turn "tug of war" game, in which the players take turns according to the outcome of a coin toss, is ∞-harmonic. This striking emergence of the ∞-Laplacian in a completely new arena supports the idea that this operator is liable to arise in many situations. This framework leads to many other operators as its ingredients are varied, and one can currently read about this in a preprint of Barron, Evans and Jensen, [14] (which is available on Evans' website as of this writing). Much more is contained in this interesting article, and many variations are derived in several different ways: many different operators, inhomogeneous equations, time dependent versions, — .

In this context, the results explained herein are quite special, corresponding, say, to merely deriving the basic properties of harmonic functions via their mean value property, and all sorts of generalizations are treated later by various theories, not using the mean value property (Poisson equations, more general elliptic operators, time dependent versions and so on). However, one should understand the Laplace equation as a starting point, and in this case, we do not know enough yet about ∞-harmonic functions.

The interesting functional equation

$$u^{\varepsilon}(x) = \frac{1}{2} \left(\max_{|y| \leq \varepsilon} u^{\varepsilon}(x + \varepsilon y) + \max_{|z| \leq \varepsilon} u^{\varepsilon}(x + \varepsilon z) \right)$$

appears in the work of [48] (see [14]). This same elegant relation plays a role in approximation arguments given in Le Gruyer [41] and Oberman [47]. Le Gruyer's work could also have been mentioned under the "metric space" heading above, and a preprint is available on arXiv, and Le Gruyer and Archer [40] is very relevant here as well.

9.5 Miscellany

We want to mention a few further papers and topics. There is the somewhat speculative offering of Evans and Yu, which ponders, among other things, the relation of the question of whether or not ∞-harmonic functions are C^1 to standard pde approaches to this question. Then there are the results of Bhattacharya [20]-[23] on some more refined properties of ∞-harmonic functions in special situations. For example, in [21], the author shows that a nonnegative ∞-harmonic function on a half-space which is continuous on the closure of the half-space is neccesarily a scalar multiple of the distance to the boundary. This result does not seem to follow in any simple way from the theory we have presented so far; it makes a more sophisticated use of consequences of the Harnack inequality, taking it up to the boundary and comparing two different functions. The results of Exercise 4.5 are very present, however, in these works.

We could go on, but it is time to stop.

References

1. ARONSSON, G., *Minimization problems for the functional* $\sup_x F(x, f(x), f'(x))$, Ark. Mat. 6 (1965), 33–53.
2. ARONSSON, G., *Minimization problems for the functional* $\sup_x F(x, f(x), f'(x))$. *II.*, Ark. Mat. 6 (1966), 409–431.
3. ARONSSON, G., *Extension of functions satisfying Lipschitz conditions*, Ark. Mat. 6 (1967), 551–561.
4. ARONSSON, G., *On the partial differential equation* $u_x{}^2 u_{xx} + 2u_x u_y u_{xy} + u_y{}^2 u_{yy} = 0$, Ark. Mat. 7 (1968), 395–425.
5. ARONSSON, G., *Minimization problems for the functional* $\sup_x F(x, f(x), f'(x))$. *III*, Ark. Mat. 7 (1969), 509–512.
6. ARONSSON, G., *On certain singular solutions of the partial differential equation* $u_x^2 u_{xx} + 2u_x u_y u_{xy} + u_y^2 u_{yy} = 0$, Manuscripta Math. 47 (1984), no. 1-3, 133–151.
7. ARONSSON, G., *Construction of singular solutions to the p-harmonic equation and its limit equation for* $p = \infty$, Manuscripta Math. 56 (1986), no. 2, 135–158.
8. ARONSSON, G., CRANDALL, M. AND JUUTINEN, P., *A tour of the theory of absolutely minimizing functions*, Bull. Amer. Math. Soc. 41 (2004), no. 4, 439–505.
9. BARLES, G. AND BUSCA, J., *Existence and comparison results for fully nonlinear degenerate elliptic equations without zeroth-order term*, Comm. Partial Diff. Equations 26 (2001), 2323–2337.
10. BARRON, E. N., *Viscosity solutions and analysis in* L^∞. Nonlinear analysis, differential equations and control (Montreal, QC, 1998), 1–60, NATO Sci. Ser. C Math. Phys. Sci., 528, Kluwer Acad. Publ., Dordrecht, 1999.
11. BARRON, E. N., JENSEN, R. R. AND WANG, C. Y., *The Euler equation and absolute minimizers of* L^∞ *functionals*, Arch. Ration. Mech. Anal. 157 (2001), no. 4, 255–283.
12. BARRON, E. N., JENSEN, R. R. AND WANG, C. Y., *Lower semicontinuity of* L^∞ *functionals*, Ann. Inst. H. Poincaré Anal. Non Linéaire 18 (2001), no. 4, 495–517.
13. BARRON, E. N. AND JENSEN, R., *Minimizing the* L^∞ *norm of the gradient with an energy constraint*, Comm. Partial Differential Equations 30 (2005) no 12, 1741–1772.

14. BARRON, E. N., EVANS, L. C. AND JENSEN, R. *The infinity Laplacian, Aronsson's equation and their generalizations*, preprint.

15. BIESKE, T., *On ∞-harmonic functions on the Heisenberg group*, Comm. Partial Differential Equations 27 (2002), no. 3-4, 727–761.

16. BIESKE, T., *Viscosity solutions on Grushin-type planes*, Illinois J. Math. 46 (2002), no. 3, 893–911.

17. BIESKE, T., *Lipschitz extensions on generalized Grushin spaces*, Michigan Math. J. 53 (2005), no. 1, 3–31.

18. BIESKE, T. AND CAPOGNA, L., *The Aronsson-Euler equation for absolutely minimizing Lipschitz extensions with respect to Carnot-Carathéodory metrics*, Trans. Amer. Math. Soc. 357 (2005), no. 2, 795–823.

19. BHATTACHARYA, T., *An elementary proof of the Harnack inequality for non-negative infinity-superharmonic functions*, Electron. J. Differential Equations, No. 44 (2001), 8 pp. (electronic).

20. BHATTACHARYA, T., *On the properties of ∞-harmonic functions and an application to capacitary convex rings*, Electron. J. Differential Equations, No. 101, (2002), 22 pp. (electronic).

21. BHATTACHARYA, T., *A note on non-negative singular infinity-harmonic functions in the half-space.*, Rev. Mat. Complut. 18 (2005), no. 2, 377–385.

22. BHATTACHARYA, T., *On the behaviour of ∞-harmonic functions near isolated points*, Nonlinear Anal. 58 (2004), no. 3-4, 333–349.

23. BHATTACHARYA, T., *On the behaviour of ∞-harmonic functions on some special unbounded domains*, Pacific J. Math. 19 (2004) no. 2, 237–253.

24. BHATTACHARYA, T., DIBENEDETTO, E. AND MANFREDI, J., *Limits as $p \to \infty$ of $\Delta_p u_p = f$ and related extremal problems*, Some topics in nonlinear PDEs (Turin, 1989). Rend. Sem. Mat. Univ. Politec. Torino 1989, Special Issue, 15–68 (1991).

25. CHAMPION, T., DE PASCALE L., *A principle of comparison with distance function for absolute minimizers*, preprint.

26. CHAMPION, T., DE PASCALE L. AND PRINARI F., *Γ-convergence and absolute minimizers for supremal functionals*, ESAIM Control Optim. Calc. Var., 10 (2004) no. 1, 14–27.

27. CRANDALL, M. G., *An efficient derivation of the Aronsson equation*, Arch. Rational Mech. Anal. 167 (2003) 271–279.

28. CRANDALL, M. G., GUNARSON, G. AND WANG, P. Y., *Uniqueness of ∞-harmonic functions in unbounded domains*, preprint.

29. CRANDALL, M. G., EVANS, L. C. AND GARIEPY, R. F., *Optimal Lipschitz extensions and the infinity Laplacian*, Calc. Var. Partial Differential Equations 13 (2001), no. 2, 123–139.

30. CRANDALL, M. G. AND EVANS, L. C., *A remark on infinity harmonic functions*, Proceedings of the USA-Chile Workshop on Nonlinear Analysis (Viã del Mar-Valparaiso, 2000), 123–129 (electronic), Electron. J. Differ. Equ. Conf., 6, Southwest Texas State Univ., San Marcos, TX, 2001.

31. CRANDALL, M. G., ISHII, H. AND LIONS, P. L., *User's guide to viscosity solutions of second-order partial differential equations*, Bull. Am. Math. Soc. 27 (1992), 1–67.

32. EVANS, L. C., ON SOLVING CERTAIN NONLINEAR PARTIAL DIFFERENTIAL EQUATIONS BY ACCRETIVE OPERATOR METHODS, Israel J. Math. 36 (1980), 225-247.

33. EVANS, L. C. AND YU, Y., *Various properties of solutions of the infinity-Laplace equation*, preprint.

34. GARIEPY, R., WANG, C. AND YU, Y. *Generalized cone comparison principle for viscosity solutions of the Aronsson equation and absolute minimizers*, preprint, 2005.

35. ISHII, HITOSHI, On the equivalence of two notions of weak solutions, viscosity solutions and distribution solutions. Funkcial. Ekvac. 38 (1995), no. 1, 101–120.

36. JENSEN, R. R., *Uniqueness of Lipschitz extensions: minimizing the sup norm of the gradient*, Arch. Rational Mech. Anal. 123 (1993), no. 1, 51–74.

37. JUUTINEN, P., *Minimization problems for Lipschitz functions via viscosity solutions*, Ann. Acad. Sci. Fenn. Math. Diss. No. 115 (1998), 53 pp.

38. JUUTINEN, P., *Absolutely minimizing Lipschitz extensions on a metric space*, Ann. Acad. Sci. Fenn. Math. 27 (2002), no. 1, 57–67.

39. JUUTINEN, PETRI; LINDQVIST, PETER; MANFREDI, JUAN J. On the equivalence of viscosity solutions and weak solutions for a quasi-linear equation. SIAM J. Math. Anal. 33 (2001), no. 3, 699–717 (electronic).

40. LE GRUYER, E. AND ARCHER, J. C., *Harmonious extensions*, SIAM J. Math. Anal. 29 (1998), no. 1, 279–292.

41. LE GRUYER, E., *On absolutely minimizing extensions and the PDE $\Delta_\infty(u) = 0$*, preprint.

42. LINDQVIST, P. AND MANFREDI, J., *The Harnack inequality for ∞-harmonic functions* Electron. J. Differential Equations (1995), No. 04, approx. 5 pp. (electronic).

43. LINDQVIST, P. AND MANFREDI, J., *Note on ∞-superharmonic functions*. Rev. Mat. Univ. Complut. Madrid 10 (1997) 471–480.

44. MCSHANE, E. J., *Extension of range of functions*, Bull. Amer. Math. Soc. 40 (1934), 837–842.

45. MIL'MAN, V. A., *Lipschitz extensions of linearly bounded functions*, (Russian) Mat. Sb. 189 (1998), no. 8, 67–92; translation in Sb. Math. 189 (1998), no. 7-8, 1179–1203.

46. MIL'MAN, V. A., *Absolutely minimal extensions of functions on metric spaces*, (Russian) Mat. Sb. 190 (1999), no. 6, 83–110; translation in Sb. Math. 190 (1999), no. 5-6, 859–885.

47. OBERMAN, A., *A convergent difference scheme for the infinity Laplacian: construction of absolutely minimizing Lipschitz extensions*, Math. Comp. 74 (2005), no. 251, 1217–1230.

48. PERES, Y., SCHRAMM, O., SHEFFIELD, S. AND WILSON, D., paper to appear

49. SAVIN, O., *C^1 regularity for infinity harmonic functions in two dimensions*, preprint.

50. WANG, C. Y., *The Aronsson equation for absolute minimizers of $L\infty$ -functionals associated with vector fields satisfying Hörmander's condition*. To appear in Trans. Amer. Math. Soc.

51. WHITNEY, H., *Analytic extensions of differentiable functions defined in closed sets*, Trans. Amer. Math. Soc. 36 (1934), no. 1, 63–89.

52. WU, Y., *Absolute minimizers in Finsler metrics*, Ph.D. dissertation, UC Berkeley, 1995.

53. YU, Y., *Viscosity solutions of Aronsson equations*, Arch. Ration. Mech. Anal., to appear

54. YU, Y., *Tangent lines of contact for the infinity Laplacian*, Calc. Var. Partial Differential Equations 21 (2004), no. 4, 349–355.

Weak KAM Theory and Partial Differential Equations

Lawrence C. Evans

Department of Mathematics
University of California,
Berkeley, CA, 94720-3840, USA
evans@math.berkeley.edu

1 Overview, KAM theory

These notes record and slightly modify my 5 lectures from the CIME conference on "Calculus of variations and nonlinear partial differential equations", held in Cetraro during the week of June 27 - July 2, 2005, organized by Bernard Dacorogna and Paolo Marcellini. I am proud to brag that this was the third CIME course I have given during the past ten years, the others at the meetings on "Viscosity solutions and applications" (Montecatini Terme, 1995) and on "Optimal transportation and applications" (Martina Franca, 2001).

My intention was (and is) to introduce some new PDE methods developed over the past 6 years in so-called "weak KAM theory", a subject pioneered by J. Mather and A. Fathi. Succinctly put, the goal of this subject is the employing of dynamical systems, variational and PDE methods to find "integrable structures" within general Hamiltonian dynamics.

My main references for most of these lectures are Fathi's forthcoming book [F5] (as well as his sequence of short notes [F1]–[F4]) and my paper [E-G1] with Diogo Gomes. A nice recent survey is Kaloshin [K]; and see also the survey paper [E6].

1.1 Classical Theory

I begin with a quick recounting of classical Lagrangian and Hamiltonian dynamics, and a brief discussion of the standard KAM theorem. A good elementary text is Percival and Richards [P-R], and Goldstein [G] is a standard reference.

The Lagrangian Viewpoint

Definition. We are given a function $L : \mathbb{R}^n \times \mathbb{R}^n \to \mathbb{R}$ called the *Lagrangian*, $L = L(v, x)$. We regard $x \in \mathbb{R}^n$, $x = (x_1, \ldots, x_n)$, as the *position* variable and $v \in \mathbb{R}^n$, $v = (v_1, \ldots, v_n)$, as the *velocity*.

Hypotheses. We hereafter assume that
(i) there exist constants $0 < \gamma \leq \Gamma$ such that

$$\Gamma|\xi|^2 \geq \sum_{i,j=1}^{n} L_{v_i v_j}(v, x)\xi_i\xi_j \geq \gamma|\xi|^2 \qquad \text{(uniform convexity)} \qquad (1.1)$$

for all ξ, v, x; and
(ii) the mapping

$$x \mapsto L(v, x) \text{ is } \mathbb{T}^n\text{-periodic} \qquad \text{(periodicity)} \qquad (1.2)$$

for all x, where $\mathbb{T}^n = [0, 1]^n$ denotes the unit cube in $\mathbb{R}^n$, with opposite faces identified.

Definition. Given a curve $\mathbf{x} : [0, T] \to \mathbb{R}$, we define its *action* to be

$$A_T[\mathbf{x}(\cdot)] := \int_0^T L(\dot{\mathbf{x}}(t), \mathbf{x}(t))\, dt$$

where $\dot{} = \frac{d}{dt}$.

Theorem 1.1 (Euler–Lagrange Equation). *Suppose that $x_0, x_T \in \mathbb{R}^n$ are given, and define the admissible class of curves*

$$\mathcal{A} := \{\mathbf{y} \in C^2([0, T]; \mathbb{R}^n) \mid \mathbf{y}(0) = x_0,\ \mathbf{y}(T) = x_T\}.$$

Suppose $\mathbf{x}(\cdot) \in \mathcal{A}$ and

$$A_T[\mathbf{x}(\cdot)] = \min_{\mathbf{y} \in \mathcal{A}} A_T[\mathbf{y}(\cdot)].$$

Then the curve $\mathbf{x}(\cdot)$ solves the Euler–Lagrange equations

(E-L) $\qquad -\dfrac{d}{dt}(D_v L(\dot{\mathbf{x}}, \mathbf{x})) + D_x L(\dot{\mathbf{x}}, \mathbf{x}) = 0 \qquad (0 \leq t \leq T).$

A Basic Example. The Lagrangian $L = \frac{|v|^2}{2} - W(x)$ is the difference between the kinetic energy $\frac{|v|^2}{2}$ and the potential energy $W(x)$. In this case the Euler-Lagrange equations read

$$\ddot{\mathbf{x}} = -DW(\mathbf{x}). \qquad \square$$

The Hamiltonian Viewpoint.

In view of the uniform convexity of $v \mapsto L(v, x)$, we can uniquely and smoothly solve the equation

$$p = D_v L(v, x)$$

for

$$v = \mathbf{v}(p, x).$$

Definition. We define the *Hamiltonian*

$$H(p, x) := p \cdot \mathbf{v}(p, x) - L(\mathbf{v}(p, x), x). \tag{1.3}$$

Equivalently,

$$H(p.x) = \max_{v \in \mathbb{R}^n} (p \cdot v - L(v, x)). \tag{1.4}$$

Theorem 1.2 (Hamiltonian Dynamics). *Suppose* $\mathbf{x}(\cdot)$ *solves the Euler-Lagrange equations* (E-L). *Define*

$$\mathbf{p}(t) := D_v L(\mathbf{x}(t), \dot{\mathbf{x}}(t)).$$

Then the pair $(\mathbf{x}(\cdot), \mathbf{p}(\cdot))$ *solves Hamilton's equations*

(H)
$$\begin{cases} \dot{\mathbf{x}} = D_P H(\mathbf{p}, \mathbf{x}) \\ \dot{\mathbf{p}} = -D_x H(\mathbf{p}, \mathbf{x}). \end{cases}$$

Also,

$$\frac{d}{dt} H(\mathbf{p}, \mathbf{x}) = 0;$$

this is conservation of energy.

Basic Example Again. For the Lagrangian $L = \frac{|v|^2}{2} - W(x)$, the corresponding Hamiltonian is $H(p, x) = \frac{|p|^2}{2} + W(x)$, the sum of the kinetic and potential energies. The Hamiltonian dynamics (H) then read

$$\begin{cases} \dot{\mathbf{x}} = \mathbf{p} \\ \dot{\mathbf{p}} = -DW(\mathbf{x}); \end{cases}$$

and the total energy is conserved. □

Canonical Changes of Variables, Generating Functions

Let $u : \mathbb{R}^n \times \mathbb{R}^n \to \mathbb{R}$, $u = u(P, x)$, be a given smooth function, called a *generating function*. Consider the formulas

$$\begin{cases} p := D_x u(P, x) \\ X := D_P u(P, x). \end{cases} \tag{1.5}$$

Assume we can solve (1.5) globally for X, P as smooth functions of x, p, and vice versa:

$$\begin{matrix} X = \mathbf{X}(p, x) \\ P = \mathbf{P}(p, x) \end{matrix} \quad \Leftrightarrow \quad \begin{matrix} x = \mathbf{x}(P, X) \\ p = \mathbf{p}(P, X). \end{matrix} \tag{1.6}$$

We next study the Hamiltonian dynamics in the new variables:

Theorem 1.3 (Change of variables). *Let* $(\mathbf{x}(\cdot), \mathbf{p}(\cdot))$ *solve Hamilton's equations* (H). *Define*

$$\mathbf{X}(t) := \mathbf{X}(\mathbf{p}(t), \mathbf{x}(t))$$
$$\mathbf{P}(t) := \mathbf{P}(\mathbf{p}(t), \mathbf{x}(t)).$$

Then

(K) $$\begin{cases} \dot{\mathbf{X}} = D_P K(\mathbf{P}, \mathbf{X}) \\ \dot{\mathbf{P}} = -D_X K(\mathbf{P}, \mathbf{X}) \end{cases}$$

for the new Hamiltonian

$$K(P, X) := H(\mathbf{p}(P, X), \mathbf{x}(P, X)).$$

Definition. A transformation $\Psi : \mathbb{R}^n \times \mathbb{R}^n \to \mathbb{R}^n \times \mathbb{R}^n$,

$$\Psi(p, x) = (P, X)$$

is called *canonical* if it preserves the Hamiltonian structure. This means

$$(D\Psi)^T J \, D\Psi = J$$

for

$$J = \begin{pmatrix} 0 & I \\ -I & 0 \end{pmatrix}$$

See any good classical mechanics textbook, such as Goldstein [G], for more explanation. The change of variables $(p, x) \mapsto (P, X)$ induced by the generating function $u = u(P, x)$ is canonical.

Hamilton–Jacobi PDE

Suppose now our generating function $u = u(P, x)$ solves the *stationary Hamilton-Jacobi equation*

$$H(D_x u, x) = \bar{H}(P), \tag{1.7}$$

where the right hand side at this point of the exposition just denotes some function of P alone. Then

$$K(P, X) = H(p, x) = H(D_x u, x) = \bar{H}(P);$$

and so (K) becomes

$$\begin{cases} \dot{\mathbf{X}} = D\bar{H}(\mathbf{P}) \\ \dot{\mathbf{P}} = 0. \end{cases} \tag{1.8}$$

The point is that these dynamics are trivial. In other words, if we can canonically change variables $(p, x) \mapsto (P, X)$ using a generating function u that solves a PDE of the form (1.7), we can easily solve the dynamics in the new variables.

Definition. We call the Hamiltonian H *integrable* if there exists a *canonical mapping*

$$\Phi : \mathbb{R}^n \times \mathbb{R}^n \to \mathbb{R}^n \times \mathbb{R}^n, \quad \Phi(P, X) = (p, x),$$

such that

$$(H \circ \Phi)(P, X) =: \bar{H}(P)$$

is a function only of P.

We call $P = (P_1, \ldots, P_n)$ the *action variables* and $X = (X_1, \ldots, X_n)$ the *angle variables*.

1.2 KAM Theory

A key question is whether we can in fact construct a canonical mapping Φ as above, converting to action-angle variables. This is in general impossible, since our PDE (1.7) will not usually have a solution u smooth in x and P; and, even if it does, it is usually not possible globally to change variables according to (1.5) and (1.6).

But KAM (Kolmogorov-Arnold-Moser) theory tells us that we can in fact carry out this procedure for a Hamiltonian that is an appropriate small perturbation of a Hamiltonian depending only on p. To be more specific, suppose our Hamiltonian H has the form

$$H(p, x) = H^0(p) + K_\varepsilon^0(p, x), \tag{1.9}$$

where we regard the term K_ε^0 as a small perturbation to the integrable Hamiltonian H^0. Assume also that $x \mapsto K_\varepsilon^0(p, x)$ is $\mathbb{T}^n$-periodic.

Generating Functions, Linearization.

We propose to find a generating function having the form

$$u(P, x) = P \cdot x + v(P, x),$$

where v is small and periodic in x. Owing to (1.5) we would then change variable through the implicit formulas

$$\begin{cases} p = D_x u = P + D_x v \\ X = D_P u = x + D_P v. \end{cases} \tag{1.10}$$

We consequently must build v so that

$$H(D_x u, x) = \bar{H}(P), \tag{1.11}$$

the expression on the right to be determined.

Now according to (1.9), (1.10) and (1.11), we want

$$H^0(P + D_x v) + K_\varepsilon^0(P + D_x v, x) = \bar{H}(P). \tag{1.12}$$

We now make the informal assumption that K_ε^0, v, and their derivatives are $O(\varepsilon)$ as $\varepsilon \to 0$. Then

$$H^0(P) + DH^0(P) \cdot D_x v + K_\varepsilon^0(P, x) = \bar{H}(P) + O(\varepsilon^2). \tag{1.13}$$

Drop $O(\varepsilon^2)$ term and write

$$\omega(P) := DH^0(P).$$

Then

$$\omega(P) \cdot D_x v + K_\varepsilon^0(P, x) = \bar{H}(P) - H^0(P); \tag{1.14}$$

this is the formal linearization of our nonlinear PDE (1.12).

Fourier series

Next we use Fourier series to try to build a solution v of (1.14). To simplify a bit, suppose for this section that $\mathbb{T}^n = [0, 2\pi]^n$ (and not $[0, 1]^n$). We can then write

$$K_\varepsilon^0(P, x) = \sum_{k \in \mathbb{Z}^n} \hat{k}(P, k) e^{ik \cdot x}$$

for the Fourier coefficients

$$\hat{k}(P, k) := \frac{1}{(2\pi)^n} \int_{\mathbb{T}^n} K_\varepsilon^0(P, x) e^{-ik \cdot x} \, dx.$$

Let us seek a solution of (1.14) having the form

$$v(P, x) = \sum_{k \in \mathbb{Z}^n} \hat{v}(P, k) e^{ik \cdot x}, \tag{1.15}$$

the Fourier coefficients $\hat{v}(P, k)$ to be selected. Plug (1.15) into (1.14):

$$i \sum_{k \in \mathbb{Z}^n} (\omega(P) \cdot k) \hat{v}(P, x) e^{ik \cdot x} + \sum_{k \in \mathbb{Z}^n} \hat{k}(P, k) e^{ik \cdot x} = \bar{H}(P) - H^0(P).$$

The various terms agree if we *define*

$$\bar{H}(P) := H^0(P) + \hat{k}(P, 0)$$

and set

$$\hat{v}(P, x) := \frac{i\hat{k}(P, k)}{\omega(P) \cdot k} \qquad (k \neq 0).$$

We have therefore derived the approximate solution

$$v(P, x) = i \sum_{k \neq 0} \frac{\hat{k}(P, k)}{\omega(P) \cdot k} e^{ik \cdot x}, \tag{1.16}$$

assuming both that $\omega(P) \cdot k \neq 0$ for all nonzero $k \in \mathbb{Z}^n$ and that the series (1.16) converges.

Small divisors

To make rigorous use the foregoing calculations, we need first to ensure that $\omega(P) \cdot k \neq 0$, with some quantitative control:

Definition. A vector $\omega \in \mathbb{R}^n$ is of type (L, γ) if

$$|k \cdot \omega| \geq \frac{L}{|k|^\gamma} \qquad \text{for all } k \in \mathbb{Z}^n, \ k \neq 0, \tag{1.17}$$

This is a "nonresonance condition". It turns out that most ω satisfy this condition for appropriate L, γ. Indeed, if $\gamma > n - 1$, then

$$|\{\omega \in B(0, R) \mid |k \cdot \omega| \leq \frac{L}{|k|^\gamma} \text{ for some } k \in \mathbb{Z}^n\}| \to 0 \quad \text{as } L \to 0.$$

Statement of KAM Theorem

We now explicitly put

$$K_\varepsilon^0(p, x) = \varepsilon K^0(p, x)$$

in (1.9), so that

$$H(p, x) = H^0(p) + \varepsilon K^0(p, x),$$

and assume

(H1)
$$\begin{cases} \text{there exists } p^* \in \mathbb{R}^n \text{ such that} \\ \quad \omega^* = DH^0(p^*) \\ \text{is type } (L, \gamma) \text{ for some } L, \gamma > 0; \end{cases}$$

(H2)
$$D^2 H^0(p^*) \text{ is invertible};$$

(H3)
$$H^0, K^0 \text{ are real analytic}.$$

Theorem 1.4 (KAM). *Under hypotheses* (H1)–(H3), *there exists* $\varepsilon_0 > 0$ *such that for each* $0 < \varepsilon \le \varepsilon_0$, *there exists* P^* *(close to* p^**) and a smooth mapping*

$$\Phi(P^*, \cdot) : \mathbb{R}^n \to \mathbb{R}^n \times \mathbb{R}^n$$

such that for each $x_0 \in \mathbb{R}^n$

$$\Phi(P^*, x_0 + t\omega^*) =: (\mathbf{x}(t), \mathbf{p}(t))$$

solves the Hamiltonian dynamics (H).

The idea of the proof is for $k = 0, 1, \ldots$, iteratively to construct $\Phi^k, P_k, H^k, \varepsilon_k$ so that $P_0 = p^*$,

$$\omega^* = D\bar{H}^k(P_k) \quad \text{for all } k;$$
$$H^{k+1} := H^k \circ \Phi^k;$$

and

$$H^k(p, x) = \bar{H}^k(p) + K^k(p, x),$$

where $\|K^k\| \le \varepsilon_k$ and the error estimates ε_k converge to 0 very rapidly. Then

$$P_k \to P^*, \quad \bar{H}^k(P_k) \to \bar{H}(P^*);$$

and

$$\Phi := \lim_{k \to \infty} \Phi^k \circ \Phi^{k-1} \circ \cdots \circ \Phi^1$$

is the required change of variables.

The full details of this procedure are very complicated. See Wayne's discussion [W] for much more explanation and for references to the vast literature on KAM.

Remark. We have

$$\begin{cases} \mathbf{X}(t) = x_0 + t\omega^* \\ \mathbf{P}(t) \equiv P^* \end{cases} \tag{1.18}$$

for

$$\omega^* = D\bar{H}(P^*)$$

But note that, in spite of my notation, Φ is really only defined for $P = P^*$.

$\square$

2 Weak KAM Theory: Lagrangian Methods

Our goal in this and the subsequent sections is extending the foregoing classical picture into the large. The resulting, so-called "weak KAM theory" is a global and nonperturbative theory (but is in truth pretty weak, at least as compared with the assertions from the previous section). There are two approaches to these issues: the Lagrangian, dynamical systems methods (discussed in this section) and the nonlinear PDE methods (explained in the next section).

The following discussion follows Albert Fathi's new book [F5], which the interested reader should consult for full details of the proofs. Related expositions are Forni–Mather [Fo-M] and Mañé [Mn].

2.1 Minimizing Trajectories

Notation. If $\mathbf{x} : [0, T] \to \mathbb{R}^n$ and

$$A_T[\mathbf{x}(\cdot)] \leq A_T[\mathbf{y}(\cdot)]$$

for all $\mathbf{y}(0) = \mathbf{x}(0)$, $\mathbf{y}(T) = \mathbf{x}(t)$, we call $\mathbf{x}(\cdot)$ a *minimizer* of the action $A_T[\cdot]$ on the time interval $[0, T]$.

Theorem 2.1 (Velocity Estimate). *For each $T > 0$, there exists a constant C_T such that*

$$\max_{0 \leq t \leq T} |\dot{\mathbf{x}}(t)| \leq C_T \tag{2.1}$$

for each minimizer $\mathbf{x}(\cdot)$ on $[0, T]$.

Idea of Proof. This is a fairly standard derivative estimate for solutions of the Euler–Lagrange equation (E-L). $\square$

2.2 Lax–Oleinik Semigroup

Definition. Let $v \in C(\mathbb{T}^n)$ and set

$$T_t^- v(x) := \inf\{v(\mathbf{x}(0)) + \int_0^t L(\dot{\mathbf{x}}, \mathbf{x}) \, ds \mid \mathbf{x}(t) = x\}$$

We call the family of nonlinear operators $\{T_t^-\}_{t \geq 0}$ the *Lax–Oleinik semigroup*.

Remark. The infimum above is attained. So in fact there exists a curve $\mathbf{x}(\cdot)$ such that $\mathbf{x}(t) = x$ and

$$T_t^- v(x) = v(\mathbf{x}(0)) + \int_0^t L(\dot{\mathbf{x}}, \mathbf{x}) \, dt. \quad \square$$

Theorem 2.2 (Regularization). *For each time $t > 0$ there exists a constant C_t such that*

$$[T_t^- v]_{Lip} \le C_t \tag{2.2}$$

for all $v \in C(\mathbb{T}^n)$.

Idea of Proof. Exploit the strict convexity of $v \mapsto L(v, x)$, as in similar arguments in Chapter 3 on my PDE text [E1]. □

We record below some properties of the Lax–Oleinik semigroup acting on the space $C(\mathbb{T}^n)$, with max norm $\| \ \|$.

Theorem 2.3 (Properties of Lax–Oleinik Semigroup).

(i) $T_t^- \circ T_s^- = T_{t+s}^-$ (semigroup property).
(ii) $v \le \hat{v}$ implies $T_t^- v \le T_t^- \hat{v}$.
(iii) $T_t^- (v + c) = T_t^- v + c$.
(iv) $\|T_t^- v - T_t^- \hat{v}\| \le \|v - \hat{v}\|$ (nonexpansiveness).
(v) *For all* $v \in C(\mathbb{T}^n)$, $\lim_{t \to 0} T_t^- v = v$ *uniformly.*
(vi) *For all* v, $t \mapsto T_t^- v$ *is uniformly continuous.*

2.3 The Weak KAM Theorem

We begin with an abstract theorem about nonlinear mappings on a Banach space X:

Theorem 2.4 (Common Fixed Points). *Suppose $\{\phi_t\}_{t \ge 0}$ is a semigroup of nonexpansive mappings of X into itself. Assume also for all $t > 0$ that $\phi_t(X)$ is precompact in X and for all $x \in X$ that $t \mapsto \phi_t(x)$ is continuous.*
   Then there exists a point x^* such that

$$\phi_t(x^*) = x^*   \text{for all times } t \ge 0.$$

Idea of Proof. Fix a time $t_0 > 0$. We will first show that ϕ_{t_0} has a fixed point.
   Let $0 < \lambda < 1$. Then there exists according to the Contraction Mapping Theorem a unique element x_λ satisfying

$$\lambda \phi_{t_0}(x_\lambda) = x_\lambda.$$

By hypothesis $\{\phi_{t_0}(x_\lambda) \mid 0 < \lambda < 1\}$ is precompact. So there exist $\lambda_j \to 1$ for which $x_{\lambda_j} \to x$ and

$$\phi_{t_0}(x) = x.$$

Now let x_n be fixed point of $\phi_{1/2^n}$. Then x_n is fixed point also of $\phi_{k/2^n}$ for $k = 1, 2, \ldots$. Using compactness, we show that $x_{n_j} \to x^*$ and that $\phi_t(x^*) = x^*$ for all $t \geq 0$. $\square$

Next is an important result of Albert Fathi [F1]:

Theorem 2.5 (Weak KAM Theorem). *There exists a function $v_- \in C(\mathbb{T}^n)$ and a constant $c \in \mathbb{R}$ such that*

$$T_t^- v_- + ct = v_- \qquad \text{for all } t \geq 0. \tag{2.3}$$

Idea of proof. We apply Theorem 2.4 above. For this, let us write for functions $v, \hat{v} \in C(\mathbb{T}^n)$

$$v \sim \hat{v}$$

if $v - \hat{v} \equiv$ constant; and define also the equivalence class

$$[v] := \{\hat{v} \mid v \sim \hat{v}\}.$$

Set

$$X := \{[v] \mid v \in C(\mathbb{T}^n)\}$$

with the norm

$$\|[v]\| := \min_{a \in \mathbb{R}} \|v + a\mathbb{1}\|,$$

where $\mathbb{1}$ denotes is the constant function identically equal to 1.

We have $T_t : X \to X$, according to property (iii) of Theorem 2.3. Hence there exists a common fixed point

$$T_t[v]^* = [v]^* \qquad (t \geq 0).$$

Selecting any representative $v_- \in [v]^*$, we see that

$$T_t v_- = v_- + c(t)$$

for some $c(t)$. The semigroup property implies $c(t + s) = c(t) + c(s)$, and consequently $c(t) = ct$ for some constant c. $\square$

2.4 Domination

Notation. Let $w \in C(\mathbb{T}^n)$. We write

$$w \prec L + c \qquad (\text{``}w \text{ is dominated by } L + c\text{''})$$

if

$$w(\mathbf{x}(b)) - w(\mathbf{x}(a)) \leq \int_a^b L(\dot{\mathbf{x}}, \mathbf{x}) ds + c(b - a) \tag{2.4}$$

for all times $a < b$ and for all Lipschitz continuous curves $\mathbf{x} : [a, b] \to \mathbb{R}^n$.

Remark.

$$w \prec L + c \text{ if and only if } w \le ct + T_t^- w \text{ for all } t \ge 0.$$

□

Theorem 2.6 (Domination and PDE).

(i) *If* $w \prec L + c$ *and the gradient* $Dw(x)$ *exists at a point* $x \in \mathbb{T}^n$, *then*

$$H(Dw(x), x) \le c. \tag{2.5}$$

(ii) *Conversely, if* w *is Lipschitz continuous and* $H(Dw, x) \le c$ *a.e., then*

$$w \prec L + c.$$

Idea of Proof. (i) Select any curve $\mathbf{x}$ with $\mathbf{x}(0) = x$, $\dot{\mathbf{x}}(0) = v$. Then

$$\frac{w(\mathbf{x}(t)) - w(x)}{t} \le \frac{1}{t} \int_0^t L(\dot{\mathbf{x}}, \mathbf{x}) \, ds + c.$$

Let $t \to 0$, to discover

$$v \cdot Dw \le L(v, x) + c;$$

and therefore

$$H(Dw(x), x) = \max_v (v \cdot Dw - L(v, x)) \le c.$$

(ii) If w is smooth, we can compute

$$w(\mathbf{x}(b)) - w(\mathbf{x}(a)) = \int_a^b \frac{d}{dt} w(\mathbf{x}(t)) \, ds$$

$$= \int_a^b Dw(\mathbf{x}) \cdot \dot{\mathbf{x}} \, dt$$

$$\le \int_a^b L(\dot{\mathbf{x}}, \mathbf{x}) + H(Dw(\mathbf{x}), \mathbf{x}) \, dt$$

$$\le \int_a^b L(\dot{\mathbf{x}}, \mathbf{x}) \, dt + c(b - a).$$

See Fathi's book [F5] for what to do when w is only Lipschitz.   □

2.5 Flow invariance, characterization of the constant c

Notation. (i) We will hereafter write

$$T(\mathbb{T}^n) := \mathbb{R}^n \times \mathbb{T}^n = \{(v, x) \mid v \in \mathbb{R}^n, \ x \in \mathbb{T}^n\}$$

for the *tangent bundle* over $\mathbb{T}^n$, and

$$T^*(\mathbb{T}^n) := \mathbb{R}^n \times \mathbb{T}^n = \{(p, x) \mid p \in \mathbb{R}^n, \ x \in \mathbb{T}^n\}$$

for the *cotangent bundle*. We work in the tangent bundle for the Lagrangian viewpoint, and in the cotangent bundle for the Hamiltonian viewpoint.

(ii) Consider this initial-value problem for the Euler-Lagrange equation:

$$\begin{cases} -\frac{d}{dt}(D_v L(\dot{\mathbf{x}}, \mathbf{x})) + D_x L(\dot{\mathbf{x}}, \mathbf{x}) = 0 \\ \mathbf{x}(0) = x, \ \dot{\mathbf{x}}(0) = v. \end{cases} \tag{2.6}$$

We define the *flow map* $\{\phi_t\}_{t \in \mathbb{R}}$ on $T(\mathbb{T}^n)$ by the formula

$$\phi_t(v, x) := (\mathbf{v}(t), \mathbf{x}(t)), \tag{2.7}$$

where $\mathbf{v}(t) = \dot{\mathbf{x}}(t)$.

Definition. A probability measure μ on the tangent bundle $T(\mathbb{T}^n)$ is *flow invariant* if

$$\int_{T(\mathbb{T}^n)} \Phi(\phi_t(v, x)) \, d\mu = \int_{T(\mathbb{T}^n)} \Phi(v, x) \, d\mu$$

for each bounded continuous function Φ.

Following is an elegant interpretation of the constant c from the Weak KAM Theorem 2.5, in terms of action minimizing flow invariant measures:

Theorem 2.7 (Characterization of c). *The constant from Theorem 2.5 is given by the formula*

$$c = -\inf \left\{ \int_{T(\mathbb{T}^n)} L(v, x) \, d\mu \mid \mu \text{ flow invariant, probability measure} \right\}. \tag{2.8}$$

Definitions. (i) The *action* of a flow-invariant measure μ is

$$A[\mu] := \int_{T(\mathbb{T}^n)} L(v, x) \, d\mu.$$

(ii) We call μ a *minimizing* (or *Mather*) *measure* if

$$-c = A[\mu]. \tag{2.9}$$

Idea of Proof. 1. Recall that

$$T_t^- v_- = v_- + ct \quad \text{and} \quad v_- \prec L + c.$$

Let us write

$$\phi_s(v, x) = (\dot{\mathbf{x}}(s), \mathbf{x}(s))$$

for $(v, x) \in T(\mathbb{T}^n)$, and put

$$\pi(v, x) := x$$

for the projection of $T(\mathbb{T}^n)$ onto $\mathbb{T}^n$.

Then

$$v_-(\pi\phi_1(v, x)) - v_-(\pi(v, x)) \le \int_0^1 L(\phi_s(v, x)) \, ds + c. \qquad (2.10)$$

Integrate with respect to a flow-invariant probability measure μ:

$$0 = \int_{T(\mathbb{T}^n)} v_-(\pi\phi_1(v, x)) - v_-(\pi(v, x)) \, d\mu$$

$$\le \int_0^1 \int_{T(\mathbb{T}^n)} L(\phi_s(v, x)) \, d\mu ds + c$$

$$= \int_{T(\mathbb{T}^n)} L(v, x) \, d\mu + c.$$

Therefore

$$-c \le \int_{T(\mathbb{T}^n)} L(v, x) \, d\mu \qquad \text{for all flow-invariant } \mu.$$

2. We must now manufacture a measure giving equality above. Fix $x \in \mathbb{T}^n$. Find a curve $\mathbf{x} : (-\infty, 0] \to \mathbb{R}^n$ such that $\mathbf{x}(0) = x$, $\dot{\mathbf{x}}(0) = v$; and for all times $t \le 0$:

$$v_-(\mathbf{x}(0)) - v_-(\mathbf{x}(t)) = \int_t^0 L(\phi_s(v, x)) \, ds - ct.$$

Define for $t \ge 0$ the measure μ_t by the rule

$$\mu_t(\Phi) = \frac{1}{t} \int_{-t}^0 \Phi(\phi_s(v, x)) \, ds$$

for each continuous function Φ. Then

$$\frac{v_-(x) - v_-(\mathbf{x}(-t))}{t} = \int_{T(\mathbb{T}^n)} L \, d\mu_t + c.$$

Sending $t_k \to -\infty$, we deduce that

$$d\mu_{t_k} \rightharpoonup d\mu$$

weakly as measures, for some flow invariant probability measure μ that satisfies

$$-c = \int_{T(\mathbb{T}^n)} L \, d\mu.$$

□

2.6 Time-reversal, Mather set

It is sometimes convenient to redo our theory with time reversed, by introducing the *backwards Lax–Oleinik semigroup*:

Definition.

$$T_t^+ v(s) := \sup \left\{ v(\mathbf{x}(t)) - \int_0^t L(\dot{\mathbf{x}}, \mathbf{x}) ds \mid \mathbf{x}(0) = x \right\}.$$

As above, there exists a function $v_+ \in C(\mathbb{T}^n)$ such that

$$T_t^+ v_+ - ct = v_+ \qquad \text{for all } t \geq 0$$

for the same constant c described in Theorem 2.7.

Definition. We define the *Mather set.*

$$\tilde{M}_0 := \overline{\bigcup_\mu \text{spt}(\mu)},$$

the union over all minimizing measures μ as above. The *projected Mather set* is

$$M_0 := \pi(\tilde{M}_0).$$

One goal of weak KAM theory is studying the structure of the Mather set (and the related Aubry set), in terms of the underlying Hamiltonian dynamics. See Fathi [F5] for much more.

3 Weak KAM Theory: Hamiltonian and PDE Methods

3.1 Hamilton–Jacobi PDE

In this section, we reinterpret the foregoing ideas in terms of the theory of *viscosity solutions* of nonlinear PDE:

Theorem 3.1 (Viscosity Solutions).

(i) *We have*

$$H(Dv_\pm, x) = c \quad a.e.$$

(ii) *In fact*

$$\begin{cases} H(Dv_-, x) = c \\ -H(Dv_+, x) = -c \end{cases} \tag{3.1}$$

in sense of viscosity solutions.

Idea of Proof. 1. There exists a minimizing curve $\mathbf{x} : (\infty, 0] \to \mathbb{R}^n$ such that $\mathbf{x}(0) = x$, $\dot{\mathbf{x}}(0) = v$, and

$$v_-(x) = v_-(\mathbf{x}(-t)) + \int_{-t}^{0} L(\dot{\mathbf{x}}, \mathbf{x})\, ds + ct \quad (t \geq 0).$$

If v is differentiable at x, then we deduce as before that

$$v \cdot Dv_-(x) = L(v, x) + c,$$

and this implies $H(Dv_-, x) \geq c$. But always $H(Dv_-, x) \leq 0$.

2. The assertions about $v_\pm$ being viscosity solutions follow as in Chapter 10 of my PDE book [E1]. □

3.2 Adding P Dependence

Motivated by the discussion in Section 1 about the classical theory of canonical transformation to action-angle variables, we next explicitly add dependence on a vector P.

So select $P \in \mathbb{R}^n$ and define the *shifted Lagrangian*

$$\hat{L}(v, x) := L(v, x) - P \cdot v. \tag{3.2}$$

The corresponding Hamiltonian is

$$\hat{H}(p, x) = \max_v (p \cdot v - \hat{L}(v, x)) = \max_v ((p + P) \cdot v - L(v, x)),$$

and so

$$\hat{H}(p, x) = H(P + p, x). \tag{3.3}$$

As above, we find a constant $c(P)$ and periodic functions $v_\pm = v_\pm(P, x)$ so that

$$\begin{cases} H(P + Dv_-, x) = \hat{H}(Dv_-, x) = c(P) \\ -H(P + Dv_+, x) = -\hat{H}(Dv_+, x) = -c(P) \end{cases}$$

in the viscosity sense.

Notation. We write

$$\bar{H}(P) := c(P), \quad u_- := P \cdot x + v_-, \quad u_+ := P \cdot x + v_+.$$

Then

$$\begin{cases} H(Du_-, x) = \bar{H}(P) \\ -H(Du_+, x) = -\bar{H}(P) \end{cases} \tag{3.4}$$

in the sense of viscosity solutions.

3.3 Lions–Papanicolaou–Varadhan Theory

A PDE construction of $\bar{H}$

We next explain an alternative, purely PDE technique, due to Lions - Papanicolaou - Varadhan [L-P-V], for finding the constant $c(P)$ as above and $v = v_-$.

Theorem 3.2 (More on viscosity solutions). *For each vector $P \in \mathbb{R}^n$, there exists a unique real number $c(P)$ for which we can find a viscosity solution of*

$$\begin{cases} H(P + D_x v, x) = c(P) \\ v \text{ is } \mathbb{T}^n\text{-periodic.} \end{cases} \tag{3.5}$$

Remark. In addition v is semiconcave, meaning that $D^2 v \le CI$ for some constant C. □

Idea of Proof. 1. *Existence.* Routine viscosity solution methods assert the existence of a unique solution v^ε of

$$\varepsilon v^\varepsilon + H(P + D_x v^\varepsilon, x) = 0.$$

Since H is periodic in x, uniqueness implies v^ε is also periodic.
   It is now not very difficult to derive the uniform estimates

$$\max |D_x v^\varepsilon|, |\varepsilon v^\varepsilon| \le C.$$

Hence we may extract subsequences for which

$$v^{\varepsilon_j} \to v \text{ uniformly}, \quad \varepsilon_j v^{\varepsilon_j} \to -c(P).$$

It is straightforward to confirm that v is a viscosity solution of $H(P + D_x v, x) = c(P)$.

   2. *Uniqueness of $c(P)$.* Suppose also

$$H(P + D_x \hat{v}, x) = \hat{c}(P).$$

We may assume that $\hat{c}(P) > c(P)$ and that $\hat{v} < v$, upon adding a constant to v, if necessary. Then

$$\delta\hat{v} + H(P + D_x\hat{v}, x) > \delta v + H(P + D_x v, x)$$

in viscosity sense, if $\delta > 0$ is small. The viscosity solution comparison principle then implies the contradiction $\hat{v} \geq v$. □

Notation. In agreement with our previous notation, we write

$$\bar{H}(P) := c(P), \quad u = P \cdot x + v, \tag{3.6}$$

and call $\bar{H}$ the *effective Hamiltonian*. Then

(E) $$H(D_x u, x) = \bar{H}(P).$$

We call (E) the *generalized eikonal* equation.

Effective Lagrangian

Dual to the effective Hamiltonian is the *effective Lagrangian*:

$$\bar{L}(V) := \max_P(P \cdot V - \bar{H}(P)) \tag{3.7}$$

for $V \in \mathbb{R}^n$.

We sometime write $\bar{L} = \bar{H}^*$ to record this Legendre transform. Then

$$P \in \partial\bar{L}(V) \quad \Leftrightarrow \quad V \in \partial\bar{H}(P) \quad \Leftrightarrow \quad P \cdot V = \bar{L}(V) + \bar{H}(P), \tag{3.8}$$

∂ denoting the (possibly multivalued) subdifferential of a convex function.

Theorem 3.3 (Characterization of effective Lagrangian). *We have the formula*

$$\bar{L}(V) = \inf\left\{ \int_{T(\mathbb{T}^n)} L(v, x)\, d\mu \mid \mu \text{ flow invariant}, \int_{T(\mathbb{T}^n)} v\, d\mu = V \right\}. \tag{3.9}$$

Compare this with Theorem 2.7.

Proof. Denote by $\tilde{L}(V)$ the right hand side of (3.9). Then

$$-\bar{H}(P) = \inf_\mu\left\{ \int_{T(\mathbb{T}^n)} L(v, x) - P \cdot v\, d\mu \right\}$$

$$= \inf_{\mu, V}\left\{ \int_{T(\mathbb{T}^n)} L(v, x)\, d\mu - P \cdot V \mid \int_{T(\mathbb{T}^n)} v\, d\mu = V \right\}$$

$$= \inf_V\{\tilde{L}(V) - P \cdot V\}$$

$$= -\sup_V\{P \cdot V - \tilde{L}(V)\}.$$

So $\tilde{L} = \bar{H}^* = \bar{L}$. □

Application: Homogenization of Nonlinear PDE

Theorem 3.4 (Homogenization). *Suppose g is bounded and uniformly continous, and u^ε is the unique bounded, uniformly continuous viscosity solution of the initial-value problem*

$$\begin{cases} u_t^\varepsilon + H(Du^\varepsilon, \frac{x}{\varepsilon}) = 0 & (t > 0) \\ u^\varepsilon = g & (t = 0). \end{cases}$$

Then $u^\varepsilon \to u$ locally uniformly, where u solves the homogenized equation

$$\begin{cases} u_t + \bar{H}(Du) = 0 & (t > 0) \\ u = g & (t = 0). \end{cases}$$

Idea of Proof. Let ϕ be a smooth function and suppose that $u - \phi$ has a strict maximum at the point (x_0, t_0). Define the perturbed test function

$$\phi^\varepsilon(x, t) := \phi(x, t) + \varepsilon v\left(\frac{x}{\varepsilon}\right),$$

where v is a periodic viscosity solution of

$$H(P + Dv, x) = \bar{H}(P)$$

for $P = D\phi(x_0, t_0)$.

Assume for the rest of the discussion that v is smooth. Then ϕ^ε is smooth and $u^\varepsilon - \phi^\varepsilon$ attains a max at a point $(x_\varepsilon, t_\varepsilon)$ near (x_0, t_0). Consequently

$$\phi_t^\varepsilon + H\left(D\phi^\varepsilon, \frac{x_\varepsilon}{\varepsilon}\right) \leq 0.$$

And then

$$\phi_t + H\left(D\phi(x_\varepsilon, t_\varepsilon) + Dv\left(\frac{x_\varepsilon}{\varepsilon}\right), \frac{x_\varepsilon}{\varepsilon}\right) \approx \phi_t + \bar{H}(D\phi(x_0, t_0)) \leq 0.$$

The reverse inequality similarly holds if $u - \phi$ has a strict minimum at the point (x_0, t_0).

See my old paper [E2] for what to do when v is not smooth.   □

3.4 More PDE Methods

In this section we apply some variational and nonlinear PDE methods to study further the structure of the Mather minimizing measures μ and viscosity solutions $u = P \cdot x + v$ of the eikonal equation (E).

Hereafter μ denotes a Mather minimizing measure in $T(\mathbb{T}^n)$. It will be more convenient to work with Hamiltonian variables, and so we define ν to be the pushforward of μ onto the contangent bundle $T^*(\mathbb{T}^n)$ under the change of variables $p = D_v L(v, x)$.

For reference later, we record these properties of ν, inherited from μ:

(A)
$$V = \int_{T^*(\mathbb{T}^n)} D_p H(v, x) \, d\nu$$

(B)
$$\bar{L}(V) = \int_{T^*(\mathbb{T}^n)} L(D_p H(p, x), x) \, d\nu$$

(C)
$$\int_{T^*(\mathbb{T}^n)} \{H, \Phi\} \, d\nu = 0 \qquad \text{for all } C^1 \text{ functions } \Phi,$$

where
$$\{H, \Phi\} := D_p H \cdot D_x \Phi - D_x H \cdot D_P \Phi.$$

is the *Poisson bracket*. Statement (A) is the definition of V, whereas statement (C) is a differential form of the flow-invariance, in the Hamiltonian variables.

Notation. We will write
$$\sigma = \pi_\# \nu$$

for the push-forward of ν onto $\mathbb{T}^n$.

Now select any $P \in \partial \bar{L}(V)$. Then as above construct a viscosity solution of the generalized eikonal PDE

(E)
$$H(D_x u, x) = \bar{H}(P)$$

for
$$u = P \cdot x + v, \ v \text{ periodic}.$$

We now study properties of u in relation to the measures σ and ν, following the paper [E-G1].

Theorem 3.5 (Regularity Properties).

(i) *The function u is differentiable for σ a.-e. point $x \in \mathbb{T}^n$.*
(ii) *We have*
$$p = D_x u \quad \nu - a.e. \tag{3.10}$$

(iii) *Furthermore,*

$$\bar{H}(P) = \int_{T^*(\mathbb{T}^n)} H(p, x) \, d\nu. \tag{3.11}$$

Remark. Compare assertion (ii) with the first of the classical formulas (1.5) □.

Idea of proof. 1. Since $p \mapsto H(p, x)$ is uniformly convex, meaning $D_p^2 H \geq \gamma I$ for some positive γ, we have

$$\beta_\varepsilon(x) + H(Du^\varepsilon, x) \leq \bar{H}(P) + O(\varepsilon) \tag{3.12}$$

for the mollified function $u^\varepsilon := \eta_\varepsilon * u$ and

$$\beta_\varepsilon(x) = \frac{\gamma}{2} \int_{\mathbb{T}^n} \eta_\varepsilon(x - y) |Du(y) - Du^\varepsilon(x)|^2 dy$$

$$\approx \frac{\gamma}{2} \fint_{B(x,\varepsilon)} |Du - (Du)_{x,\varepsilon}|^2 dy.$$

In this formula $(Du)_{x,\varepsilon}$ denotes the average of Du over the ball $B(x, \varepsilon)$.

2. Uniform convexity also implies

$$\frac{\gamma}{2} \int_{T^*(\mathbb{T}^n)} |Du^\varepsilon - p|^2 d\nu \leq \int_{T^*(\mathbb{T}^n)} H(Du^\varepsilon, x) - H(p, x) - D_p H(p, x) \cdot (Du^\varepsilon - p) \, d\nu.$$

Now $Du^\varepsilon = P + Dv^\varepsilon$ and

$$\int_{T^*(\mathbb{T}^n)} D_p H \cdot Dv^\varepsilon \, d\nu = 0 \qquad \text{by property (C)}.$$

So (3.12) implies

$$\frac{\gamma}{2} \int_{T^*(\mathbb{T}^n)} |Du^\varepsilon - p|^2 d\nu \leq \int_{T^*(\mathbb{T}^n)} \bar{H}(P) - H(p, x) - D_p H(p, x) \cdot (P - p) - \beta_\varepsilon(x) \, d\nu$$

$$+ O(\varepsilon)$$

$$\leq \bar{H}(P) - V \cdot P - \int_{T^*(\mathbb{T}^n)} \bar{H}(p, x) - D_p H(p, x) \cdot p - \beta_\varepsilon(x) \, d\nu$$

$$+ O(\varepsilon)$$

$$= \bar{L}(V) - \int_{T^*(\mathbb{T}^n)} L(D_p H, x) - \beta_\varepsilon(x) \, d\nu + O(\varepsilon),$$

according to (3.8). Consequently

$$\int_{T^*(\mathbb{T}^n)} |Du^\varepsilon - p|^2 \, d\nu + \int_{\mathbb{T}^n} \beta_\varepsilon(x) \, d\sigma \leq O(\varepsilon). \tag{3.13}$$

Recall also that u is semiconcave. Therefore (3.13) implies σ-a.e. point is Lebesgue point of Du, and so Du exists σ-a.e. It then follows from (3.13) that $Du^\varepsilon \to Du$ ν-a.e.

2. We have $\int_{T^*(\mathbb{T}^n)} H(p, x) \, d\nu = \int_{\mathbb{T}^n} H(Du, x) \, d\sigma = \bar{H}(P)$.   □

Remark. In view of (3.10), the flow invariance condition (C) implies

$$\int_{\mathbb{T}^n} D_p H(Du, x) \cdot D\phi \, d\sigma = 0$$

for all $\phi \in C^1(\mathbb{T}^n)$; and so the measure σ is a weak solution of the *generalized transport* (or *continuity*) *equation*

(T) $$\mathrm{div}(\sigma D_p H(D_x u, x)) = 0.   □$$

3.5 Estimates

Now we illustrate how our two key PDE, the generalized eikonal equation (E) for u and the generalized transport equation (T) for σ and u, together yield more information about the smoothness of u.

To simplify the presentation, we take the standard example

$$H(p, x) = \frac{1}{2}|p|^2 + W(x)$$

for this section. Then our eikonal and transport equations become

(E) $$\frac{1}{2}|Du|^2 + W(x) = \bar{H}(P)$$

(T) $$\mathrm{div}(\sigma Du) = 0.$$

If u is smooth, we could differentiate (E) twice with respect to x_k:

$$\sum_{i=1}^{n} u_{x_i} u_{x_i x_k x_k} + u_{x_i x_k} u_{x_i x_k} + W_{x_k x_k} = 0,$$

sum on k, and then integrate with respect to σ over $\mathbb{T}^n$:

$$\int_{\mathbb{T}^n} D_x u \cdot D_x(\Delta u) \, d\sigma + \int_{\mathbb{T}^n} |D_x^2 u|^2 d\sigma = -\int_{\mathbb{T}^n} \Delta W \, d\sigma.$$

According to (T) the first term equals zero.

This establishes the formal estimate

$$\int_{\mathbb{T}^n} |D_x^2 u|^2 \, d\sigma \leq C;$$ (3.14)

and a related rigorous estimate involving difference quotients holds if u is not smooth: see [E-G1].

Reworking the proof using appropriate cutoff functions, we can derive as well the formal bound

$$|D_x^2 u|^2 \leq C \quad \sigma \text{ a.e.;} \tag{3.15}$$

and a related rigorous estimate involving difference quotients is valid if u is not smooth. Again, see [E-G1] for the details. We thereby establish the inequality

$$|Du(y) - Du(x)| \leq C|x - y| \quad \text{for } x \in \text{spt}(\sigma), \text{ a.e. } y \in \mathbb{T}^n. \tag{3.16}$$

In particular, even though Du may be multivalued, we can bound

$$\text{diam}(Du(y)) \leq C \text{ dist}(y, \text{spt}(\sigma))$$

for some constant C. This is a sort of quantitative estimate on how far the support of σ lies from the "shocks" of the gradient of u.

An application of these estimates is a new proof of Mather's regularity theorem for the support of the minimizing measures:

Theorem 3.6 (Mather). *The support of μ lies on a Lipschitz graph in $T(\mathbb{T}^n)$, and the support of ν lies on a Lipschitz graph in $T^*(\mathbb{T}^n)$.*

Remark. In addition, if u is smooth in x and P, we have the formal bound

$$\int_{\mathbb{T}^n} |D_{xP}^2 u|^2 \, d\sigma \leq C. \tag{3.17}$$

A related rigorous estimate involving difference quotients holds if u is not smooth. As an application, we show in [E-G1] that if $\bar{H}$ is twice differentiable at P, then

$$|D\bar{H}(P) \cdot \xi| \leq C(\xi \cdot D^2 \bar{H}(P)\xi)^{1/2}.$$

for all vectors ξ and some constant C.   □

4 An Alternative Variational/PDE Construction

4.1 A new Variational Formulation

This section follows [E3], to discuss an alternate variational/PDE technique for discovering the structure of weak KAM theory.

A Minimax Formula

Our motivation comes from "the calculus of variations in the sup-norm", as presented for instance in Barron [B]. We start with the following observation, due to several authors:

Theorem 4.1 (Minimax Formula for $\bar{H}$). *We have*

$$\bar{H}(P) = \inf_{v \in C^1(\mathbb{T}^n)} \max_{x \in \mathbb{T}^n} H(P + Dv, x). \tag{4.1}$$

Idea of Proof. Write $u = P \cdot x + v$, where u is our viscosity solution of (E), and put $\hat{u} = P \cdot x + \hat{v}$, where $\hat{v}$ is any C^1, periodic function. Then convexity implies

$$H(Du, x) + D_p H(Du, x) \cdot (D\hat{u} - Du) \le H(D\hat{u}, x).$$

Integrate with respect to σ:

$$\bar{H}(P) + \int_{\mathbb{T}^n} D_p H(Du, x) \cdot D(\hat{v} - v) \, d\sigma \le \int_{\mathbb{T}^n} H(D\hat{u}, x) \, d\sigma$$
$$\le \max_x H(P + D\hat{v}, x).$$

Therefore (T) implies that

$$\bar{H}(P) \le \inf_{\hat{v}} \max_x H(P + D\hat{v}, x).$$

Furthermore, if $v^\varepsilon := \eta^\varepsilon * v$, where η^ε is a standard mollifier, then

$$H(P + Dv^\varepsilon, x) \le \bar{H}(P) + O(\varepsilon);$$

and consequently

$$\max_x H(P + Dv^\varepsilon, x) \le \bar{H}(P) + O(\varepsilon). \quad \Box$$

Remark. See Fathi–Siconolfi [F-S1] for the construction of a C^1 subsolution. $\quad \Box$

A New Variational Setting

The minimax formula provided by Theorem 4.1 suggests that we may be able somehow to approximate the "max" above by an exponential integral with a large parameter.

To be precise, we fix a large constant k and then introduce the integral functional

$$I_k[v] := \int_{\mathbb{T}^n} e^{kH(P + Dv, x)} \, dx.$$

(My paper [E3] cites some relevant references, and see also Marcellini [M1], [M2] and Mascolo–Migliorini [M-M] for more about such problems with exponential growth.)

Let v_k be the unique minimizer among perodic functions, subject to the normalization that

$$\int_{\mathbb{T}^n} v^k \, dx = 0.$$

As usual, put $u^k := P \cdot x + v^k$. The Euler–Lagrange equation then reads

$$\operatorname{div}(e^{kH(Du^k,x)} D_p H(Du^k, x)) = 0. \tag{4.2}$$

Passing to Limits.

We propose to study the asymptotic limit of the PDE (4.2) as $k \to \infty$. We will discover that the structure of weak KAM theory appears in the limit.

Notation. It will be convenient to introduce the normalizations

$$\sigma^k := \frac{e^{kH(Du^k,x)}}{\int_{\mathbb{T}^n} e^{kH(Du^k,x)} dx} \tag{4.3}$$

and

$$\bar{H}^k(P) := \frac{1}{k} \log \left(\int_{\mathbb{T}^n} e^{kH(Du^k,x)} dx \right). \tag{4.4}$$

Define also

$$d\mu^k := \delta_{\{v=D_p H(Du^k,x)\}} \sigma^k \, dx.$$

Passing as necessary to subsequences, we may assume

$$\sigma^k dx \rightharpoonup d\sigma \quad \text{weakly as measures on } \mathbb{T}^n,$$
$$d\mu^k \rightharpoonup d\mu \quad \text{weakly as measures on } T(\mathbb{T}^n),$$
$$u^k \to u \quad \text{uniformly.}$$

A main question is what PDE (if any) do u and σ satisfy?

Theorem 4.2 (Weak KAM in the Limit).
   (i) *We have*
$$\bar{H}(P) = \lim_{k \to \infty} \bar{H}^k(P). \tag{4.5}$$

   (ii) *The measure μ is a Mather minimizing measure.*
   (iii) *Furthermore*

$$\begin{cases} H(Du, x) \leq \bar{H}(P) & a.e. \\ H(Du, x) = \bar{H}(P) & \sigma\text{-}a.e. \end{cases} \tag{4.6}$$

(iv) *The measure σ is a weak solution of*

$$\text{div}(\sigma D_p H(Du, x)) = 0. \tag{4.7}$$

(v) *In addition, u is a viscosity solution of* Aronsson's PDE

$$-A_H[u] := -\sum_{i,j=1}^{n} H_{p_i} H_{p_j} u_{x_i x_j} - \sum_{i=1}^{n} H_{x_i} H_{p_i} = 0. \tag{4.8}$$

4.2 Application: Nonresonance and Averaging.

We illustrate some uses of the approximation (4.2) by noting first that our (unique) solution $u^k := P \cdot x + v^k$ is smooth in both x and P. We can therefore legitimately differentiate in both variables, and are encouraged to do so by the classical formulas (1.5).

Derivatives of $\bar{H}^k$

We can also calculate the first and second derivatives in P of the smooth approximate effective Hamiltonian $\bar{H}^k$. Indeed, a direct computation (cf. [E3]) establishes the useful formulas

$$D\bar{H}^k(P) = \int_{\mathbb{T}^n} D_p H(Du^k, x) \, d\sigma^k \tag{4.9}$$

and

$$D^2\bar{H}^k(P) = k \int_{\mathbb{T}^n} (D_p H(Du^k, x) D_{xP}^2 u^k - D\bar{H}^k(P)) \otimes (D_p H D_{xP}^2 u^k - D\bar{H}) \, d\sigma^k$$
$$+ \int_{\mathbb{T}^n} D_p^2 H(Du^k, x) D_{xP}^2 u \otimes D_{xP}^2 u \, d\sigma^k, \tag{4.10}$$

where for notational convenience we write

$$d\sigma^k := \sigma^k dx.$$

Nonresonance

Next, let us assume $\bar{H}$ is differentiable at P and that $V = D\bar{H}(P)$ satisfies the weak nonresonance condition

$$V \cdot m \neq 0 \quad \text{for all } m \in \mathbb{Z}^n, \ m \neq 0. \tag{4.11}$$

What does this imply about the limits as $k \to \infty$ of σ^k and u^k??

Theorem 4.3 (Nonresonance and Averaging). *Suppose* (4.11) *holds and also*

$$|D^2\bar{H}^k(P)| \leq C$$

for all large k.
     Then

$$\lim_{k\to\infty} \int_{\mathbb{T}^n} \Phi(D_P u^k)\sigma^k \, dx = \int_{\mathbb{T}^n} \Phi \, dX. \qquad (4.12)$$

for all continuous, periodic functions $\Phi = \Phi(X)$.

Interpretation. As discussed in Section 1, if we could really change to the action-angle variables X, P according to (1.5) and (1.6), we would obtain the linear dynamics

$$\mathbf{X}(t) = X_0 + tV = X_0 + tD\bar{H}(P).$$

In view of the nonresonance condition (4.11), it follows then that

$$\lim_{T\to\infty} \frac{1}{T} \int_0^T \Phi(\mathbf{X}(t)) \, dt = \int_{\mathbb{T}^n} \Phi(X) \, dX \qquad (4.13)$$

for all continuous, periodic functions Φ.

Observe next from (4.2) that σ^k is a solution of the transport equation

$$\operatorname{div}(\sigma^k D_p H(Du^k, x)) = 0. \qquad (4.14)$$

Now a direct calculation shows that if u is a smooth solution of (E), then

$$\hat{\sigma} := \det D^2_{xP} u$$

in fact solves the same transport PDE:

$$\operatorname{div}(\hat{\sigma} D_p H(Du, x)) = 0.$$

This suggests that maybe we somehow have $\sigma^k \approx \det D^2_{xP} u^k$ in the asymptotic limit $k \to \infty$. Since

$$\int_{\mathbb{T}^n} \Phi(D_P u)\hat{\sigma} \, dx = \int_{\mathbb{T}^n} \Phi(X) \, dX$$

if the mapping $X = D_P u(P, x)$ is one-to-one and onto, we might then hope that something like (4.12) is valid. □

Idea of Proof. Recall (4.14) and note also that

$$w := e^{2\pi i m \cdot D_P u^k} = e^{2\pi i m \cdot (x + D_P v^k)}$$

is periodic (even though $D_P u^k = x + D_P v^k$ is not). Hence

$$0 = \int_{\mathbb{T}^n} D_p H(Du^k, x) \cdot D_x w \sigma^k \, dx$$

$$= 2\pi i \int_{\mathbb{T}^n} e^{2\pi i m \cdot D_P u^k} m \cdot D_p H D^2_{xP} u^k \sigma^k \, dx$$

$$= 2\pi i \int_{\mathbb{T}^n} e^{2\pi i m \cdot D_P u^k} m \cdot D\bar{H}^k(P) \sigma^k \, dx$$

$$+ 2\pi i \int_{\mathbb{T}^n} e^{2\pi i m \cdot D_P u^k} m (D_p H D_{xP} u^k - D\bar{H}^k) \sigma^k \, dx$$

$$=: A + B.$$

Using formula (4.10), we can estimate

$$|B| \le C(\frac{1}{k}|D^2 \bar{H}^k(P)|)^{1/2} = o(1) \qquad \text{as } k \to \infty.$$

Consequently $A \to 0$. Since

$$m \cdot D\bar{H}^k(P) \to m \cdot V \neq 0,$$

we deduce that

$$\int_{\mathbb{T}^n} e^{2\pi i m \cdot D_P u^k} \sigma^k \, dx \to 0.$$

This proves (4.12) for all finite trigonometric polynomials Φ, and then, by the density of such trig polynomials, for all continuous Φ. □

5 Some Other Viewpoints and Open Questions

This concluding section collects together some comments about other work on, and extending, weak KAM theory and about possible future progress.

● **Geometric properties of the effective Hamiltonian**
    Concordel in [C1], [C2] initiated the systematic study of the geometric properties of the effective Hamiltonian $\bar{H}$, but many questions are still open.
    Consider, say, the basic example $H(p, x) = \frac{|p|^2}{2} + W(x)$ and ask how the geometric properties of the periodic potential W influence the geometric properties of $\bar{H}$, and vice versa. For example, if we know that $\bar{H}$ has a "flat spot" at its minimum, what does this imply about W?
    It would be interesting to have some more careful numerical studies here, as for instance in Gomes-Oberman [G-O].

• Nonresonance

Given the importance of nonresonance assumptions for the perturbative classical KAM theory, it is a critically important task to understand consequences of this condition for weak KAM in the large. Theorem 4.3 is a tiny step towards understanding this fundamental issue. See also Gomes [G1], [G2] for some more developments.

• Aubry and Mather sets

See Fathi's book [F5] for a more detailed discussion of the Mather set (only mentioned above), the larger Aubry set, and their dynamical systems interpretations. One fundamental question is just how, and if, Mather sets can act as global replacements for the classical KAM invariant tori.

• Weak KAM and mass transport

Bernard and Buffoni [B-B1], [B-B2], [B-B3] have rigorously worked out some of the fascinating interconnections between weak KAM theory and optimal mass transport theory (see Ambrosio [Am] and Villani [V]). Some formal relationships are sketched in my expository paper [E5].

• Stochastic and quantum analogs

My paper [E4] discusses the prospects of finding some sort of quantum version of weak KAM, meaning ideally to understand possible connections with solutions of Schrödinger's equation in the semiclassical limit $h \to 0$. This all of course sounds good, but it is currently quite unclear if any nontrivial such connections really exist.

N. Anantharaman's interesting paper [An] shows that a natural approximation scheme (independently proposed in [E4]) gives rise to Mather minimizing measures which additionally extremize an entropy functional.

Gomes in [G3] and Iturriaga and Sanchez-Morgado in [I-SM] have discussed stochastic versions of weak KAM theory, but here too many key questions are open.

• Nonconvex Hamiltonians

It is, I think, very significant that the theory [L-P-V] of Lions, Papanicolaou and Varadhan leads to the existence of solutions to the generalized eikonal equation (E) even if the Hamiltonian H is nonconvex in the momenta p: all that is really needed is the coercivity condition that $\lim_{|p| \to \infty} H(p, x) = \infty$, uniformly for $x \in \mathbb{T}^n$. In this case it remains a major problem to interpret $\bar{H}$ in terms of dynamics.

Fathi and Siconolfi [F-S2] have made great progress here, constructing much of the previously discussed theory under the hypothesis that $p \mapsto H(p, x)$ be *geometrically quasiconvex*, meaning that for each real number λ and $x \in \mathbb{T}^n$, the sublevel set $\{p \mid H(p, x) \leq \lambda\}$ is convex.

The case of Hamiltonians which are coercive, but nonconvex and nonquasiconvex in p, is completely open.

- **Aronsson's PDE.**

I mention in closing one final mystery: does Aronsson's PDE (4.8) have anything whatsoever to do with the Hamiltonian dynamics? This highly degenerate, highly nonlinear elliptic equation occurs quite naturally from the variational construction in Section 4, but to my knowledge has no interpretation in terms of dynamical systems. (See Yu [Y] and Fathi–Siconolfi [F-S3] for more on this strange PDE.)

References

[Am]   L. Ambrosio, Lecture notes on optimal transport problems, in *Mathematical Aspects of Evolving Interfaces* 1–52, Lecture Notes in Mathematics 1812, Springer, 2003.

[An]   N. Anantharaman, Gibbs measures and semiclassical approximation to action-minimizing measures, Transactions AMS, to appear.

[B]   E. N. Barron, Viscosity solutions and analysis in L^∞, in *Nonlinear Analysis, Differential Equations and Control*, Dordrecht, 1999.

[B-B1]   P. Bernard and B. Buffoni, The Monge problem for supercritical Mañé potentials on compact manifolds, to appear.

[B-B2]   P. Bernard and B. Buffoni, Optimal mass transportation and Mather theory, to appear.

[B-B3]   P. Bernard and B. Buffoni, The Mather-Fathi duality as limit of optimal transportation problems, to appear.

[C1]   M. Concordel, Periodic homogenization of Hamilton-Jacobi equations I: additive eigenvalues and variational formula, Indiana Univ. Math. J. 45 (1996), 1095–1117.

[C2]   M. Concordel, Periodic homogenization of Hamilton-Jacobi equations II: eikonal equations, Proc. Roy. Soc. Edinburgh 127 (1997), 665–689.

[E1]   L. C. Evans, *Partial Differential Equations*, American Mathematical Society, 1998. Third printing, 2002.

[E2]   L. C. Evans, Periodic homogenization of certain fully nonlinear PDE, Proc Royal Society Edinburgh 120 (1992), 245–265.

[E3]   L. C. Evans, Some new PDE methods for weak KAM theory, Calculus of Variations and Partial Differential Equations, 17 (2003), 159–177.

[E4]   L. C. Evans, Towards a quantum analogue of weak KAM theory, Communications in Mathematical Physics, 244 (2004), 311–334.

[E5]   L. C. Evans, Three singular variational problems, in *Viscosity Solutions of Differential Equations and Related Topics*, Research Institute for the Mathematical Sciences, RIMS Kokyuroku 1323, 2003.

[E6]   L. C. Evans, A survey of partial differential equations methods in weak KAM theory, Communications in Pure and Applied Mathematics 57 (2004), 445–480.

[E-G1]   L. C. Evans and D. Gomes, Effective Hamiltonians and averaging for Hamiltonian dynamics I, Archive Rational Mech and Analysis 157 (2001), 1–33.

[E-G2]   L. C. Evans and D. Gomes, Effective Hamiltonians and averaging for Hamiltonian dynamics II, Archive Rational Mech and Analysis 161 (2002), 271–305.

[F1]   A. Fathi, Théorème KAM faible et theorie de Mather sur les systemes lagrangiens, C. R. Acad. Sci. Paris Sr. I Math. 324 (1997), 1043–1046.

[F2]   A. Fathi, Solutions KAM faibles conjuguees et barrieres de Peierls, C. R. Acad. Sci. Paris Sr. I Math. 325 (1997), 649–652.

[F3]   A. Fathi, Orbites heteroclines et ensemble de Peierls, C. R. Acad. Sci. Paris Sr. I Math. 326 (1998), 1213–1216.

[F4]   A. Fathi, Sur la convergence du semi-groupe de Lax-Oleinik, C. R. Acad. Sci. Paris Sr. I Math. 327 (1998), 267–270.

[F5]   A. Fathi, *The Weak KAM Theorem in Lagrangian Dynamics* (Cambridge Studies in Advanced Mathematics), to appear.

[F-S1]   A. Fathi and A. Siconolfi, Existence of C^1 critical subsolutions of the Hamilton-Jacobi equation, Invent. Math. 155 (2004), 363–388.

[F-S2]   A. Fathi and A. Siconolfi, PDE aspects of Aubry-Mather theory for quasi-convex Hamiltonians, Calculus of Variations and PDE 22 (2005), 185–228.

[F-S3]   A. Fathi and A. Siconolfi, Existence of solutions for the Aronsson–Euler equation, to appear.

[Fo-M]   G. Forni and J. Mather, Action minimizing orbits in Hamiltonian systems, in *Transition to Chaos in Classical and Quantum Mechanics*, Lecture Notes in Math 1589, edited by S. Graffi, Springer, 1994.

[G]   H. Goldstein, *Classical mechanics* (2nd ed), Addison-Wesley, 1980.

[G1]   D. Gomes, Viscosity solutions of Hamilton-Jacobi equations and asymptotics for Hamiltonian systems, Calculus of Variations and PDE 14 (2002), 345–357.

[G2]   D. Gomes, Perturbation theory for Hamilton-Jacobi equations and stability of Aubry-Mather sets, SIAM J. Math. Analysis 35 (2003), 135–147.

[G3]   D. Gomes, A stochastic analog of Aubry-Mather theory, Nonlinearity 10 (2002), 271-305.

[G-O]   D. Gomes and A. Oberman, Computing the effective Hamiltonian using a variational approach, SIAM J. Control Optim. 43 (2004), 792–812.

[I-SM]   R. Iturriaga and H. Sanchez-Morgado, On the stochastic Aubry-Mather theory, to appear.

[K]   V. Kaloshin, Mather theory, weak KAM and viscosity solutions of Hamilton-Jacobi PDE, preprint.

[L-P-V]   P.-L. Lions, G. Papanicolaou, and S. R. S. Varadhan, Homogenization of Hamilton–Jacobi equations, unpublished, circa 1988.

[Mn]   R. Mañé, *Global Variational Methods in Conservative Dynamics*, Instituto de Matemática Pura e Aplicada, Rio de Janeiro.

[M1]   P. Marcellini, Regularity for some scalar variational problems under general growth conditions, J Optimization Theory and Applications 90 (1996), 161–181.

[M2]   P. Marcellini, Everywhere regularity for a class of elliptic systems without growth conditions, Annali Scuola Normale di Pisa 23 (1996), 1–25.

[M-M]   E. Mascolo and A. P. Migliorini, Everywhere regularity for vectorial functionals with general growth, ESAIM: Control, Optimization and Calculus of Variations 9 (2003), 399-418.

[Mt1]   J. Mather, Minimal measures, Comment. Math Helvetici 64 (1989), 375–394.

[Mt2]   J. Mather, Action minimizing invariant measures for positive definite Lagrangian systems, Math. Zeitschrift 207 (1991), 169–207.

[P-R]   I. Percival and D. Richards, *Introduction to Dynamics*, Cambridge University Press, 1982.

[V]   C. Villani, *Topics in Optimal Transportation*, American Math Society, 2003.

[W]   C. E. Wayne, An introduction to KAM theory, in *Dynamical Systems and Probabilistic Methods in Partial Differential Equations*, 3–29, Lectures in Applied Math 31, American Math Society, 1996.

[Y]   Y. Yu, L^∞ *Variational Problems, Aronsson Equations and Weak KAM Theory*, thesis, University of California, Berkeley, 2005.

Geometrical Aspects of Symmetrization

Nicola Fusco

Dipartimento di Matematica e Applicazioni
Università di Napoli "Federico II" Via Cintia, 80126 Napoli, Italy
n.fusco@unina.it

1 Sets of finite perimeter

Symmetrization is one of the most powerful mathematical tools with several applications both in Analysis and Geometry. Probably the most remarkable application of Steiner symmetrization of sets is the De Giorgi proof (see [14], [25]) of the isoperimetric property of the sphere, while the spherical symmetrization of functions has several applications to PDEs and Calculus of Variations and to integral inequalities of Poincaré and Sobolev type (see for instance [23], [24], [19], [20]).

The two model functionals that we shall consider in the sequel are: the perimeter of a set E in $\mathbb{R}^n$ and the Dirichlet integral of a scalar function u. It is well known that on replacing E or u by its Steiner symmetral or its spherical symmetrization, respectively, both these quantities decrease. This fact is classical when E is a smooth open set and u is a C^1 function ([22], [21]). Moreover, on approximating a set of finite perimeter with smooth open sets or a Sobolev function by C^1 functions, these inequalities can be easily extended by lower semicontinuity to the general setting ([19], [25], [2], [4]). However, an approximation argument gives no information about the equality case. Thus, if one is interested in understanding when equality occurs, one has to carry on a deeper analysis, based on fine properties of sets of finite perimeter and Sobolev functions.

Let us start by recalling what the Steiner symmetrization of a measurable set E is. For simplicity, and without loss of generality, in the sequel we shall always consider the symmetrization of E in the vertical direction. To this aim, it is convenient to denote the points x in $\mathbb{R}^n$ also by (x', y), where $x' \in \mathbb{R}^{n-1}$ and $y \in \mathbb{R}$. Thus, given $x' \in \mathbb{R}^{n-1}$, we shall denote by $E_{x'}$ the corresponding one-dimensional section of E

$$E_{x'} = \{y \in \mathbb{R} : (x', y) \in E\} \ .$$

The *distribution function* μ of E is defined by setting for all $x' \in \mathbb{R}^{n-1}$

$$\mu(x') = \mathcal{L}^1(E_{x'}).$$

Here and in the sequel we denote by $\mathcal{L}^k$ the Lebesgue measure in $\mathbb{R}^k$. Then, denoting the *essential projection* of E by $\pi(E)^+ = \{x' \in \mathbb{R}^{n-1} : \mu(x') > 0\}$, the *Steiner symmetral* of E with respect to the hyperplane $\{y = 0\}$ is the set

$$E^s = \{(x', y) : x' \in \pi(E)^+, |y| < \mu(x')/2\}.$$

Notice that by Fubini's theorem we get immediately that μ is a $\mathcal{L}^{n-1}$-measurable function in $\mathbb{R}^{n-1}$, hence E^s is a measurable set in $\mathbb{R}^n$ and $\mathcal{L}^n(E) = \mathcal{L}^n(E^s)$. Moreover, it is not hard to see that the diameter of E decreases under Steiner symmetrization, i.e., $\operatorname{diam}(E^s) \le \operatorname{diam}(E)$, an inequality which in turn implies ([1, Proposition 2.52]) the well known *isodiametric inequality*

$$\mathcal{L}^n(E) \le \omega_n \left(\frac{\operatorname{diam}(E)}{2}\right)^n,$$

where ω_n denotes the measure of the unit ball in $\mathbb{R}^n$.

Denoting by $P(E)$ the *perimeter* of a measurable set in $\mathbb{R}^n$, the following result states that the perimeter too decreases under Steiner symmetrization.

Theorem 1.1. *Let $E \subset \mathbb{R}^n$ be a measurable set. Then,*

$$P(E^s) \le P(E). \tag{1.1}$$

As we said before, inequality (1.1) is classic when E is a smooth set and can be proved by a simple approximation argument in the general case of a set of finite perimeter. However, following [9], we shall give here a different proof of Theorem 1.1, which has the advantage of providing valuable information in the case when (1.1) reduces to an equality.

Let us now recall the definition of perimeter. If E is a measurable set in $\mathbb{R}^n$ and $\Omega \subset \mathbb{R}^n$ is an open set, we say that E is a set of *finite perimeter* in Ω if the distributional derivative of the characteristic function of E, $D\chi_E$, is a vector-valued Radon measure in Ω, with finite total variation $|D\chi_E|(\Omega)$. Thus, denoting by $(D_1\chi_E, \dots, D_n\chi_E)$ the components of $D\chi_E$, we have that for all $i = 1, \dots, n$ and all test functions $\varphi \in C_0^1(\Omega)$

$$\int_\Omega \chi_E(x) \frac{\partial \varphi}{\partial x_i}(x)\, dx = -\int_\Omega \varphi(x)\, dD_i\chi_E(x). \tag{1.2}$$

From this formula it follows that the total variation of $D\chi_E$ in Ω can be expressed as

$$|D\chi_E|(\Omega) = \sup\left\{\sum_{i=1}^n \int_\Omega \psi_i(x)\, dD_i\chi_E : \psi \in C_0^1(\Omega; \mathbb{R}^n), \|\psi\|_\infty \le 1\right\} \tag{1.3}$$

$$= \sup\left\{\int_E \operatorname{div}\psi(x)\, dx : \psi \in C_0^1(\Omega; \mathbb{R}^n), \|\psi\|_\infty \le 1\right\}.$$

Notice that, if E is a smooth bounded open set, equation (1.2) reduces to

$$\int_{E\cap\Omega} \frac{\partial\varphi}{\partial x_i}(x)\,dx = -\int_{\partial E\cap\Omega} \varphi(x)\nu_i^E(x)\,d\mathcal{H}^{n-1}(x)\,,$$

where ν^E denotes the inner normal to the boundary of E. Here and in the sequel $\mathcal{H}^k$, $1 \le k \le n-1$, stands for the Hausdorff k-dimensional measure in $\mathbb{R}^n$. Thus, for a smooth set E,

$$D_i\chi_E = \nu_i^E \mathcal{H}^{n-1}\llcorner\partial E \qquad\qquad i=1,\ldots,n\,,$$

$$|D\chi_E|(\Omega) = \mathcal{H}^{n-1}(\partial E\cap\Omega)\,.$$

Last equation suggests to define the *perimeter of E in Ω* by setting $P(E;\Omega) = |D\chi_E|(\Omega)$. More generally, if $B \subset \Omega$ is any Borel subset of Ω, we set

$$P(E;B) = |D\chi_E|(B)\,.$$

If $\Omega = \mathbb{R}^n$, the perimeter of E in $\mathbb{R}^n$ will be denoted simply by $P(E)$. Notice that the last supremum in (1.3) makes sense for any measurable set E. Indeed, if for some E that supremum is finite, then an application of Riesz's theorem on functionals on $C_0(\Omega;\mathbb{R}^n)$ yields that $D\chi_E$ is a Radon measure and (1.3) holds. Thus, we may set for any measurable set $E \subset \mathbb{R}^n$ and any open set Ω

$$P(E;\Omega) = \sup\left\{\int_E \operatorname{div}\psi(x)\,dx : \psi \in C_0^1(\Omega;\mathbb{R}^n),\ \|\psi\|_\infty \le 1\right\}. \tag{1.4}$$

Clearly E a set of finite perimeter according to the definition given above if and only if the right hand side of (1.4) is finite. Notice also that from (1.4) it follows that if E_h is a sequence of measurable sets converging locally in measure to E in Ω, i.e., such that $\chi_{E_h} \to \chi_E$ in $L^1_{\text{loc}}(\Omega)$, then

$$P(E;\Omega) \le \liminf_{h\to\infty} P(E_h;\Omega)\,. \tag{1.5}$$

Another immediate consequence of the definition of perimeter is that $P(E;\Omega)$ does not change if we modify E by a set of zero Lebesgue measure. Moreover, it is straightforward to check that

$$P(E;\Omega\setminus\partial E) = |D\chi_E|(\Omega\setminus\partial E) = 0\,,$$

i.e., $D\chi_E$ is concentrated on the topological boundary of E.

Next example shows that in general $D\chi_E$ may be concentrated on a much smaller set. Let us denote by $B_r(x)$ the ball with center x and radius r and set $E = \cup_{i=1}^\infty B_{1/2^i}(q_i)$, where $\{q_i\}$ a dense sequence in $\mathbb{R}^n$. Then E is an open set of finite measure such that $\mathcal{L}^n(\partial E) = \infty$. However, E is a set of finite perimeter in $\mathbb{R}^n$. In fact, given $\psi \in C_0^1(\mathbb{R}^n;\mathbb{R}^n)$ with $\|\psi\|_\infty \le 1$, by applying the classical divergence theorem to the Lipschitz open sets $E_k = \cup_{i=1}^{k} B_{1/2^i}(q_i)$, with $k \geq 1$, we have

$$\int_E \operatorname{div}\psi \, dx = \lim_{k\to\infty} \int_{E_k} \operatorname{div}\psi \, dx = -\lim_{k\to\infty} \int_{\partial E_k} \langle \psi, \nu^{E_k} \rangle \, d\mathcal{H}^{n-1}$$

$$\leq \lim_{k\to\infty} \mathcal{H}^{n-1}(\partial E_k) \leq \sum_{i=1}^{\infty} \frac{n\omega_n}{2^{i(n-1)}} < \infty \, .$$

To identify the set of points where the measure "perimeter" $P(E; \cdot)$ is concentrated, we may use the Besicovitch derivation theorem (see [1, Theorem 2.22]), which guarantees that if E is a set of finite perimeter in $\mathbb{R}^n$, then for $|D\chi_E|$-a.e. point $x \in \operatorname{supp}|D\chi_E|$ (the support of the total variation of $D\chi_E$) there exists the derivative of $D\chi_E$ with respect to $|D\chi_E|$,

$$\lim_{r\to 0} \frac{D\chi_E(B_r(x))}{|D\chi_E|(B_r(x))} = \nu^E(x) \, , \tag{1.6}$$

and that

$$|\nu^E(x)| = 1 \, . \tag{1.7}$$

The set of points where (1.6) and (1.7) hold is called the *reduced boundary of E* and denoted by $\partial^* E$. If $x \in \partial^* E$, $\nu^E(x)$ is called the *generalized inner normal to E at x*. Since from Besicovitch theorem we have that $D\chi_E = \nu^E |D\chi_E| \llcorner \partial^* E$, formula (1.2) can be written as

$$\int_E \frac{\partial \varphi}{\partial x_i} \, dx = -\int_{\partial^* E} \varphi \nu_i^E \, d|D\chi_E| \qquad \text{for all } \varphi \in C_0^1(\mathbb{R}^n) \text{ and } i = 1, \ldots, n \, . \tag{1.8}$$

The following theorem ([13] or [1, Theorem 3.59]) describes the structure of the reduced boundary of a set of finite perimeter.

Theorem 1.2 (De Giorgi). *Let $E \subset \mathbb{R}^n$ be a set of finite perimeter in $\mathbb{R}^n$, $n \geq 2$. Then*

(i)
$$\partial^* E = \bigcup_{h=1}^{\infty} K_h \cup N_0 \, ,$$

where each K_h is a compact subset of a C^1 manifold M_h and $\mathcal{H}^{n-1}(N_0) = 0$;

(ii)
$$|D\chi_E| = \mathcal{H}^{n-1} \llcorner \partial^* E \, ;$$

for $\mathcal{H}^{n-1}$-a.e. $x \in K_h$, $\nu^E(x)$ is orthogonal to the tangent plane to M_h at x.

From this theorem it is clear that for a set of finite perimeter the reduced boundary plays the same role of the topological boundary for smooth sets. In particular, the integration by parts formula (1.8) becomes

$$\int_E \frac{\partial \varphi}{\partial x_i} dx = - \int_{\partial^* E} \varphi \nu_i^E \, d\mathcal{H}^{n-1} \qquad \text{for all } \varphi \in C_0^1(\mathbb{R}^n) \text{ and } i = 1, \ldots, n,$$

(1.9)

an equation very similar to the one we have when E is a smooth open set.

In the one-dimensional case sets of finite perimeter are completely characterized by the following result (see [1, Proposition 3.52]).

Proposition 1.1. *Let $E \subset \mathbb{R}$ be a measurable set. Then E has finite perimeter in $\mathbb{R}$ if and only if there exist $-\infty \leq a_1 < b_1 < a_2 < \cdots < b_{N-1} < a_N < b_N \leq +\infty$, such that E is equivalent to $\bigcup_{i=1}^{N}(a_i, b_i)$. Moreover, if Ω is an open set in $\mathbb{R}$,*

$$P(E; \Omega) = \#\{i : a_i \in \Omega\} + \#\{i : b_i \in \Omega\}.$$

Notice that from this characterization we have that if $E \subset \mathbb{R}$ is a set of finite perimeter with finite measure, then $P(E) \geq 2$. Moreover, $P(E) = 2$ if and only if E is equivalent to a bounded interval. Notice also that Proposition 1.1 yields immediately Theorem 1.1 and the characterization of the equality case in (1.1).

If we translate Theorem 1.2 in the language of Geometric Measure theory, then assertion (i) says that the reduced boundary $\partial^* E$ of a set of finite perimeter E in $\mathbb{R}^n$ is a countably $\mathcal{H}^{n-1}$-rectifiable set (see [1, Definition 2.57]), while (iii) states that for $\mathcal{H}^{n-1}$-a.e. $x \in \partial^* E$ the approximate tangent plane to $\partial^* E$ at x (see [1, Section 2.11]) is orthogonal to $\nu^E(x)$. Therefore, from the coarea formula for rectifiable sets ([1, Remark 2.94]), we get that if $g : \mathbb{R}^n \to [0, +\infty]$ is a Borel function, then

$$\int_{\partial^* E} g(x)|\nu_n(x)| \, d\mathcal{H}^{n-1}(x) = \int_{\mathbb{R}^{n-1}} dx' \int_{(\partial^* E)_{x'}} g(x', y) \, d\mathcal{H}^0(y), \qquad (1.10)$$

where $\mathcal{H}^0$ denotes the counting measure.

Setting $V = \{x \in \partial E^* : \nu_n^E(x) = 0\}$ and $g(x) = \chi_V(x)$, from (1.10) we get that

$$\int_{\mathbb{R}^{n-1}} dx' \int_{(\partial^* E)_{x'}} \chi_V(x', y) \, d\mathcal{H}^0(y) = \int_{\partial^* E} \chi_V(x)|\nu_n(x)| \, d\mathcal{H}^{n-1}(x) = 0.$$

Therefore, if E is a set of finite perimeter, then $V_{x'} = \emptyset$ for $\mathcal{L}^{n-1}$-a.e. $x' \in \mathbb{R}^{n-1}$, i.e.,

for $\mathcal{L}^{n-1}$-a.e. $x' \in \mathbb{R}^{n-1}$, $\nu_n^E(x', y) \neq 0$ for all y such that $(x', y) \in \partial^* E$.

(1.11)

Let Ω be an open subset of $\mathbb{R}^n$ and $u \in L^1(\Omega)$. We say that u is a *function of bounded variation* (shortly, a *BV-function*) in Ω, if the distributional derivative Du is a vector-valued Radon measure in Ω with finite total variation. Thus, denoting by $D_i u$, $i = 1, \ldots, n$, the components of Du we have that

$$\int_\Omega u \frac{\partial \varphi}{\partial x_i} \, dx = - \int_\Omega \varphi \, dD_i u \qquad \text{for all } \varphi \in C_0^1(\Omega). \qquad (1.12)$$

The space of functions of bounded variation in Ω will be denoted by $BV(\Omega)$. Notice that if $u \in BV(\Omega)$, then, as in (1.3), we have

$$|Du|(\Omega) = \sup \left\{ \sum_{i=1}^n \int_\Omega \psi_i(x) \, dD_i u : \ \psi \in C_0^1(\Omega; \mathbb{R}^n), \ \|\psi\|_\infty \leq 1 \right\}$$

$$= \sup \left\{ \int_\Omega u \operatorname{div}\psi(x) \, dx : \ \psi \in C_0^1(\Omega; \mathbb{R}^n), \ \|\psi\|_\infty \leq 1 \right\}.$$

Moreover, it is clear that if E is a measurable set such that $\mathcal{L}^n(E \cap \Omega) < \infty$, then $\chi_E \in BV(\Omega)$ if and only if E has finite perimeter in Ω.

In the sequel, we shall denote by $D^a u$ the absolutely continuous part of Du with respect to Lebesgue measure $\mathcal{L}^n$. The singular part of Du will be denoted by $D^s u$. Moreover, we shall use the symbol ∇u to denote the density of $D^a u$ with respect to $\mathcal{L}^n$. Therefore,

$$Du = \nabla u \mathcal{L}^n + D^s u.$$

Notice also that a function $u \in BV(\Omega)$ belongs to $W^{1,1}(\Omega)$ if and only if Du is absolutely continuous with respect to $\mathcal{L}^n$, i.e., $|Du|(B) = 0$ for all Borel sets $B \subset \Omega$ such that $\mathcal{L}^n(B) = 0$. In this case, the density of Du with respect to $\mathcal{L}^n$ reduces to the usual weak gradient ∇u of a Sobolev function.

Next result is an essential tool for studying the behavior of Steiner symmetrization with respect to perimeter.

Lemma 1.1. *Let E a set of finite perimeter in $\mathbb{R}^n$ with finite measure. Then $\mu \in BV(\mathbb{R}^{n-1})$ and for any bounded Borel function $\varphi : \mathbb{R}^{n-1} \to \mathbb{R}$*

$$\int_{\mathbb{R}^{n-1}} \varphi(x') \, dD_i \mu(x') = \int_{\partial^* E} \varphi(x') \nu_i^E(x) \, d\mathcal{H}^{n-1}(x), \qquad i = 1, \dots, n-1. \qquad (1.13)$$

Moreover, for any Borel set $B \subset \mathbb{R}^{n-1}$,

$$|D\mu|(B) \leq P(E; B \times \mathbb{R}). \qquad (1.14)$$

Proof. Let us fix $\varphi \in C_0^1(\mathbb{R}^{n-1})$ and a sequence $\{\psi_j\}$ of $C_0^1(\mathbb{R})$ functions, such that $0 \leq \psi_j(y) \leq 1$ for all $y \in \mathbb{R}$ and $j \in \mathbb{N}$, with $\lim_{j \to \infty} \psi_j(y) = 1$ for all y. For any $i \in \{1, \dots, n-1\}$, from Fubini's theorem and formula (1.9), we get immediately

$$\int_{\mathbb{R}^{n-1}} \frac{\partial \varphi}{\partial x_i}(x') \mu(x') \, dx' = \int_{\mathbb{R}^{n-1}} dx' \int_{\mathbb{R}} \frac{\partial \varphi}{\partial x_i}(x') \chi_E(x', y) \, dy$$

$$= \lim_{j \to \infty} \int_E \frac{\partial \varphi}{\partial x_i}(x') \psi_j(y) \, dx' dy \qquad (1.15)$$

$$= - \lim_{j \to \infty} \int_{\partial^* E} \varphi(x') \psi_j(y) \nu_i^E(x) \, d\mathcal{H}^{n-1}$$

$$= - \int_{\partial^* E} \varphi(x') \nu_i^E(x) \, d\mathcal{H}^{n-1}.$$

This proves that the distributional derivatives of μ are real measures with bounded variation. Therefore, since $\mathcal{L}^n(E) < \infty$, hence $\mu \in L^1(\mathbb{R}^{n-1})$, we have that $\mu \in BV(\mathbb{R}^{n-1})$ and thus, by applying (1.12) to μ, from (1.15) we get in particular that (1.13) holds with $\varphi \in C_0^1(\mathbb{R}^{n-1})$. The case of a bounded Borel function φ then follows easily by approximation (see [9, Lemma 3.1]).

Finally, when B is an open set of $\mathbb{R}^{n-1}$, (1.14) follows immediately from (1.13) and (ii) of Theorem 1.2. Again, the general case of a Borel set $B \subset \mathbb{R}^{n-1}$, follows by approximation. $\square$

Next result provides a first estimate of the perimeter of E^s. Notice that in the statement below we have to assume that E^s is a set of finite perimeter, a fact that will be proved later.

Lemma 1.2. *Let E be any set of finite perimeter in $\mathbb{R}^n$ with finite measure. If E^s is a set of finite perimeter, then*

$$P(E^s; B \times \mathbb{R}) \leq P(E; B \times \mathbb{R}) + |D_n \chi_{E^s}|(B \times \mathbb{R}) \qquad (1.16)$$

for every Borel set $B \subset \mathbb{R}^{n-1}$.

Proof. Since $\mu \in BV(\mathbb{R}^{n-1})$, by a well known property of BV functions (see [1, Theorem 3.9], we may find a sequence $\{\mu_j\}$ of nonnegative functions from $C_0^1(\mathbb{R}^{n-1})$ such that $\mu_j \to \mu$ in $L^1(\mathbb{R}^{n-1})$, $\mu_j(x') \to \mu(x')$ for $\mathcal{L}^{n-1}$-a.e. x' in $\mathbb{R}^{n-1}$, $|D\mu_j|(\mathbb{R}^{n-1}) \to |D\mu|(\mathbb{R}^{n-1})$ and $|D\mu_j| \to |D\mu|$ weakly* in the sense of measures. Then, setting

$$E_j^s = \{(x', y) \in \mathbb{R}^{n-1} \times \mathbb{R} : \mu_j(x') > 0, |y| < \mu_j(x')/2\},$$

we easily get that $\chi_{E_j^s}(x) \to \chi_{E^s}(x)$ in $L^1(\mathbb{R}^n)$. Fix an open set $U \subset \mathbb{R}^{n-1}$ and $\psi \in C_0^1(U \times \mathbb{R}, \mathbb{R}^n)$. Then, Fubini's theorem and a standard differentiation of integrals yield

$$\int_{U \times \mathbb{R}} \chi_{E_j^s} \operatorname{div}\psi \, dx = \int_U dx' \int_{-\mu_j(x')/2}^{\mu_j(x')/2} \sum_{i=1}^{n-1} \frac{\partial \psi_i}{\partial x_i} \, dy + \int_{\Omega \times \mathbb{R}} \chi_{E_j^s} \frac{\partial \psi_n}{\partial y} \, dx$$

$$= -\frac{1}{2} \int_{\pi(\operatorname{supp}\psi)} \sum_{i=1}^{n-1} \left[\psi_i\left(x', \frac{\mu_j(x')}{2}\right) - \psi_i\left(x', -\frac{\mu_j(x')}{2}\right) \right] \frac{\partial \mu_j}{\partial x_i} \, dx'$$

$$+ \int_{\Omega \times \mathbb{R}} \chi_{E_j^s} \frac{\partial \psi_n}{\partial y} \, dx,$$

where $\pi : \mathbb{R}^n \to \mathbb{R}^{n-1}$ denotes the projection over the first $n-1$ components. Thus

$$\int_{U \times \mathbb{R}} \chi_{E_j^s} \operatorname{div}\psi \, dx \leq$$

$$\leq \int_{\pi(\operatorname{supp}\psi)} \sqrt{\sum_{i=1}^{n-1} \left[\frac{1}{2}\left(\psi_i\left(x', \frac{\mu_j(x')}{2}\right) - \psi_i\left(x', -\frac{\mu_j(x')}{2}\right)\right) \right]^2} |\nabla \mu_j| \, dx' +$$

$$+ \int_{U \times \mathbb{R}} \chi_{E_j^s} \frac{\partial \psi_n}{\partial y} \, dx. \qquad (1.17)$$

If $\|\psi\|_\infty \leq 1$, from (1.17) we get

$$\int_{U \times \mathbb{R}} \chi_{E_j^s} \mathrm{div}\psi \, dx \leq \int_{\pi(\mathrm{supp}\psi)} |\nabla \mu_j| \, dx' + \int_{U \times \mathbb{R}} \chi_{E_j^s} \frac{\partial \psi_n}{\partial y} \, dx. \qquad (1.18)$$

Since $\chi_{E_j^s} \to \chi_{E^s}$ in $\mathcal{L}^1(\mathbb{R}^{n-1})$ and $\pi(\mathrm{supp}\psi)$ is a compact subset of U, recalling that $|D\mu_j| \to |D\mu|$ weakly* in the sense of measure and taking the lim sup in (1.18) as $j \to \infty$, we get

$$\begin{aligned}
\int_{U \times \mathbb{R}} \chi_{E^s} \mathrm{div}\psi \, dx &\leq |D\mu|(\pi(\mathrm{supp}\psi)) + \int_{U \times \mathbb{R}} \chi_{E^s} \frac{\partial \psi_n}{\partial y} \, dx \\
&\leq |D\mu|(U) + |D_n \chi_{E^s}|(U \times \mathbb{R}) \qquad (1.19) \\
&\leq P(E; U \times \mathbb{R}) + |D_n \chi_{E^s}|(U \times \mathbb{R}),
\end{aligned}$$

where the last inequality follows from (1.14). Inequality (1.19) implies that (1.16) holds whenever B is an open set, and hence also when B is any Borel set.   □

Remark 1.1. Notice that the argument used in the proof of Lemma 1.2 above yields that if E is a bounded set of finite perimeter, then E^s is a set of finite perimeter too. In fact, in this case, by applying (1.18) with $U = \mathbb{R}^{n-1}$ and $\|\psi\|_\infty \leq 1$ we get

$$\int_{\mathbb{R}^n} \chi_{E_j^s} \mathrm{div}\psi \, dx \leq$$
$$\leq \int_{\mathbb{R}^{n-1}} |\nabla \mu_j| \, dx' + \int_{\mathbb{R}^{n-1}} \left[\psi_n(x', \mu_j(x')/2) - \psi_n(x', -\mu_j(x')/2) \right] dx'.$$

Hence, passing to the limit as $j \to \infty$, we get, from (1.14) and the assumption that E is bounded,

$$\int_{\mathbb{R}^n} \chi_{E^s} \mathrm{div}\psi \, dx \leq |D\mu|(\mathbb{R}^{n-1}) + \int_{\mathbb{R}^{n-1}} \left[\psi_n(x', \mu(x')/2) - \psi_n(x', -\mu(x')/2) \right] dx'$$
$$\leq P(E) + 2\mathcal{L}^{n-1}(\pi(E)^+) < \infty.$$

Next result, due to Vol'pert ([26], [1, Theorem 3.108]), states that for $\mathcal{L}^{n-1}$-a.e. x' the section $E_{x'}$ is equivalent to a finite union of open intervals whose endpoints belong to the corresponding section $(\partial^* E)_{x'}$ of the reduced boundary.

Theorem 1.3. *Let E be a set of finite perimeter in $\mathbb{R}^n$. Then, for $\mathcal{L}^{n-1}$-a.e. $x' \in \mathbb{R}^{n-1}$,*

(i) $E_{x'}$ *has finite perimeter in* $\mathbb{R}$;

(ii) $\partial^* E_{x'} = (\partial^* E)_{x'}$;

(iii) $\qquad\qquad \nu_n^E(x', y) \neq 0$ *for all y such that $(x', y) \in \partial^* E$*;

(iv) $\chi_E(x', \cdot)$ *coincides $\mathcal{L}^1$-a.e. with a function $g_{x'}$ such that for all $y \in \partial^* E_{x'}$*

$$
\begin{cases}
\lim\limits_{z \to y^+} g_{x'}(z) = 1, & \lim\limits_{z \to y^-} g_{x'}(z) = 0 \quad \text{if } \nu_n^E(x', y) > 0, \\
\lim\limits_{z \to y^+} g_{x'}(z) = 0, & \lim\limits_{z \to y^-} g_{x'}(z) = 1 \quad \text{if } \nu_n^E(x', y) < 0.
\end{cases}
$$

The meaning of (i) and (ii) is clear. Property (iii) states that the section $(\partial^* E)_{x'}$ of the reduced boundary contains no vertical parts. As we have observed in (1.11), this is a consequence of the coarea formula (1.10). Finally, (iv) states that the normal $\nu^E(x)$ at a point $x \in \partial^* E$ has a positive vertical component if and only if $E_{x'}$ lies locally above x.

Notice also that from (ii) it follows that $(\partial^* E)_{x'} = \emptyset$ for $\mathcal{L}^{n-1}$-a.e. $x' \notin \pi(E)^+$ and that there exists a Borel set $G_E \subset \pi(E)^+$ such that

$$
\text{the conclusions (i)-(iv) of Theorem 1.3 hold for every} \atop x' \in G_E, \qquad \mathcal{L}^{n-1}(\pi(E)^+ \setminus G_E) = 0. \tag{1.20}
$$

Let us now give a useful representation formula for the absolutely continuous part of the gradient of μ.

Lemma 1.3. *Let $E \subset \mathbb{R}^n$ be a set of finite perimeter with finite measure. Then, for $\mathcal{L}^{n-1}$-a.e. $x' \in \pi(E)^+$,*

$$
\frac{\partial \mu}{\partial x_i}(x') = \sum_{y \in \partial^* E_{x'}} \frac{\nu_i^E(x', y)}{|\nu_n^E(x', y)|}, \qquad i = 1, \ldots, n-1. \tag{1.21}
$$

Proof. Let G_E be a Borel set satisfying (1.20) and g any function in $C_0(\mathbb{R}^{n-1})$. Set $\varphi(x') = g(x')\chi_{G_E}(x')$. From (1.13) and (1.10), recalling also (iii) and (ii) of Theorem 1.3, we have

$$
\int_{G_E} g(x')\, dD_i\mu = \int_{\partial^* E} g(x')\chi_{G_E}(x')\nu_i^E(x)\, d\mathcal{H}^{n-1}(x) =
$$

$$
= \int_{\partial^* E} g(x')\chi_{G_E}(x') \frac{\nu_i^E(x)}{|\nu_n^E(x)|}|\nu_n^E(x)|\, d\mathcal{H}^{n-1}(x)
$$

$$
= \int_{G_E} g(x') \sum_{y \in \partial^* E_{x'}} \frac{\nu_i^E(x', y)}{|\nu_n^E(x', y)|}\, dx'.
$$

Thus from this equality we get that

$$
D_i\mu \llcorner G_E = \left(\sum_{y \in \partial^* E_{x'}} \frac{\nu_i^E(x', y)}{|\nu_n^E(x', y)|} \right) \mathcal{L}^{n-1} \llcorner G_E.
$$

Hence the assertion follows, since by (1.20) $\mathcal{L}^{n-1}(\pi(E)^+ \setminus G_E) = 0$. $\qquad \square$

Remark 1.2. If E^s is a set of finite perimeter, since E and E^s have the same distribution function μ, we may apply Lemma 1.3 thus getting

$$\frac{\partial \mu}{\partial x_i}(x') = 2\frac{\nu_i^{E^s}(x', \frac{1}{2}\mu(x'))}{|\nu_n^{E^s}(x', \frac{1}{2}\mu(x'))|} \qquad \text{for } \mathcal{L}^{n-1}\text{-a.e. } x' \in \pi(E)^+. \qquad (1.22)$$

2 Steiner Symmetrization of Sets of Finite Perimeter

Let us start by proving the following version of Theorem 1.1.

Theorem 1.1 (Local version) *Let $E \subset \mathbb{R}^n$ be a set of finite perimeter, $n \geq 2$. Then E^s is also of finite perimeter and for every Borel set $B \subset \mathbb{R}^{n-1}$,*

$$P(E^s; B) \leq P(E; B). \qquad (2.1)$$

Proof. Let $E \subset \mathbb{R}^n$ be a set of finite perimeter. If $\mathcal{L}^n(E) = \infty$, by the isoperimetric inequality (3.6) below, $\mathbb{R}^n \setminus E$ has finite measure, hence $\mathcal{L}^1(\mathbb{R} \setminus E_{x'}) < \infty$ for $\mathcal{L}^{n-1}$-a.e. $x' \in \mathbb{R}^{n-1}$, $E^s = \mathbb{R}^n$ and the assertion follows trivially.

Thus we may assume that E has finite measure. For the moment, let us assume also that E^s is a set of finite perimeter (we shall prove this fact later). Let us set $G = G_E \cap G_{E^s}$, where G_E and G_{E^s} are defined as in (1.20). To prove inequality (2.1) it is enough to assume $B \subset G$ or $B \subset \mathbb{R}^{n-1} \setminus G$.

In the first case, using Theorem 1.2 (ii), Theorem 1.3 (iii), coarea formula (1.10) and formulas (1.22) and (1.21), we get easily

$$
\begin{aligned}
P(E^s; B \times \mathbb{R}) &= \int_{\partial^* E^s \cap (B \times \mathbb{R})} \frac{1}{|\nu_n^{E^s}|} |\nu_n^{E^s}| \, d\mathcal{H}^{n-1} = \int_B \sum_{y \in \partial^* E^s_{x'}} \frac{1}{|\nu_n^{E^s}(x', y)|} \, dx' \\
&= 2 \int_B \frac{1}{|\nu_n^{E^s}(x', \frac{1}{2}\mu(x'))|} \, dx' \qquad (2.2) \\
&= 2 \int_B \sqrt{1 + \sum_{i=1}^{n-1} \left(\frac{|\nu_i^{E^s}(x', \frac{1}{2}\mu(x'))|}{|\nu_n^{E^s}(x', \frac{1}{2}\mu(x'))|} \right)^2} \, dx' \\
&= \int_B \sqrt{4 + \sum_{i=1}^{n-1} \left(\frac{\partial \mu}{\partial x_i}(x') \right)^2} \, dx' \\
&= \int_B \sqrt{4 + \sum_{i=1}^{n-1} \left(\sum_{y \in \partial^* E_{x'}} \frac{\nu_i^E(x', y)}{|\nu_n^E(x', y)|} \right)^2} \, dx'.
\end{aligned}
$$

Notice that, since E has finite measure, for a.e. $x' \in \mathbb{R}^{n-1}$, $\mathcal{L}^1(E_{x'}) < \infty$ and thus $P(E_{x'}) \geq 2$. Hence from the equality above, using the discrete Minkowski inequality, we get

$$P(E^s; B \times \mathbb{R}) = \int_B \sqrt{4 + \sum_{i=1}^{n-1} \left(\sum_{y \in \partial^* E_{x'}} \frac{\nu_i^E(x', y)}{|\nu_n^E(x', y)|} \right)^2} \, dx' \qquad (2.3)$$

$$\leq \int_B \sqrt{\left(\#\{y : y \in \partial^* E_{x'}\} \right)^2 + \sum_{i=1}^{n-1} \left(\sum_{y \in \partial^* E_{x'}} \frac{\nu_i^E(x', y)}{|\nu_n^E(x', y)|} \right)^2} \, dx'$$

$$\leq \int_B \sum_{y \in \partial^* E_{x'}} \sqrt{1 + \sum_{i=1}^{n-1} \left(\frac{\nu_i^E(x', y)}{|\nu_n^E(x', y)|} \right)^2} \, dx'$$

$$= \int_B \sum_{y \in \partial^* E_{x'}} \frac{1}{|\nu_n^E(x', y)|} \, dx' = P(E; B \times \mathbb{R}),$$

where the last two equalities, as in (2.2), are a consequence of the coarea formula and of the assumption $B \subset G_E$.

When $B \subset \mathbb{R}^{n-1} \setminus G$, we use (1.6), Theorem 1.2 (ii), coarea formula again, Theorem 1.3 (ii) and the fact that $\mathcal{L}^{n-1}(\pi(E)^+ \cap B) = 0$, thus getting

$$|D_n \chi_{E^s}|(B \times \mathbb{R}) = \int_{\partial^* E^s \cap (B \times \mathbb{R})} |\nu_n^{E^s}| d\mathcal{H}^{n-1} = \int_B \#\{y \in \partial^* E_{x'}^s\} dx'$$

$$= \int_{B \setminus \pi(E)^+} \#\{y \in \partial^* E_{x'}^s\} dx' = 0,$$

where the last equality is a consequence of the fact that $E_{x'}^s = \emptyset$ for all $x' \notin \pi(E)^+$. Then (2.1) immediately follows from (1.16).

Let us now prove now that E^s is a set of finite perimeter. If E is bounded, this property follows from what we have already observed in Remark 1.1. If E is not bounded, we may always find a sequence of smooth bounded open sets E_h such that $\mathcal{L}^n(E \Delta E_h) \to 0$ and $P(E_h) \to P(E)$ as $h \to \infty$ (see [1, Theorem 3.42]). Notice that, by Fubini's theorem,

$$\mathcal{L}^n(E^s \Delta (E_h)^s) = \int_{\mathbb{R}^{n-1}} |\mathcal{L}^1(E_{x'}^s) - \mathcal{L}^1((E_h)_{x'}^s)| \, dx'$$

$$= \int_{\mathbb{R}^{n-1}} |\mathcal{L}^1(E_{x'}) - \mathcal{L}^1((E_h)_{x'})| \, dx'$$

$$\leq \int_{\mathbb{R}^{n-1}} |\mathcal{L}^1(E_{x'} \Delta (E_h)_{x'})| \, dx' = \mathcal{L}^n(E \Delta E_h).$$

Therefore, from the lower semicontinuity of perimeters with respect to convergence in measure (1.5) and from what we have proved above we get

$$P(E^s) \leq \liminf_{h \to \infty} P((E_h)^s) \leq \lim_{h \to \infty} P(E_h) = P(E)$$

and thus E^s has finite perimeter. $\square$

The result we have just proved was more or less already known in the literature though with a different proof (see for instance [25]). The interesting point of the above proof is that it provides almost immediately some non trivial information about the case when equality holds in (1.1), as shown by the next result.

Theorem 2.1. *Let E be a set of finite perimeter in $\mathbb{R}^n$, with $n \geq 2$, such that equality holds in (1.1). Then, either E is equivalent to $\mathbb{R}^n$ or $\mathcal{L}^n(E) < \infty$ and for $\mathcal{L}^{n-1}$-a.e. $x' \in \pi(E)^+$*

$$E_{x'} \text{ is equivalent to a segment } (y_1(x'), y_2(x')), \tag{2.4}$$

$$(\nu_1^E, \ldots, \nu_{n-1}^E, \nu_n^E)(x', y_1(x')) = (\nu_1^E, \ldots, \nu_{n-1}^E, -\nu_n^E)(x', y_2(x')). \tag{2.5}$$

Proof. If $\mathcal{L}^n(E) = \infty$, as we have already observed in the previous proof, $E^s = \mathbb{R}^n$. Then, since $P(E) = P(E^s) = 0$, it follows that also E is equivalent to $\mathbb{R}^n$.

If $\mathcal{L}^n(E) < \infty$, from the assumption $P(E) = P(E^s)$ and from inequality (2.1) it follows that $P(E^s; B \times \mathbb{R}) = P(E; B \times \mathbb{R})$ for all Borel sets $B \subset \mathbb{R}^{n-1}$. By applying this equality with $B = G$, where G is the set introduced in the proof above, it follows that both inequalities in (2.3) are indeed equalities. In particular, since the first inequality holds as an equality, we get

$$\#\{y : y \in \partial^* E_{x'}\} = 2 \qquad \text{for } \mathcal{L}^{n-1}\text{-a.e. } x' \in G.$$

Hence (2.4) follows, recalling that, by (1.20), $\mathcal{L}^{n-1}(\pi(E)^+ \setminus G) = 0$.
The fact that also the second inequality in (2.3) is an equality implies that

$$\frac{\nu_i^E(x', y_1(x'))}{|\nu_n^E(x', y_1(x'))|} = \frac{\nu_i^E(x', y_2(x'))}{|\nu_n^E(x', y_2(x'))|}$$

for $i = 1, \ldots, n-1$ and for $\mathcal{L}^{n-1}$-a.e. $x' \in G$.

From this equation, since $|\nu^E| = 1$, we have that $\nu_i^E(x', y_1(x')) = \nu_i^E(x', y_2(x'))$ and $|\nu_n^E(x', y_1(x'))| = |\nu_n^E(x', y_2(x'))|$ for $\mathcal{L}^{n-1}$-a.e. $x' \in G$. Then, equality $\nu_n^E(x', y_1(x')) = -\nu_n^E(x', y_2(x'))$ is an easy consequence of assertion (iv) of Theorem 1.3. Hence, (2.5) follows. $\square$

As we have just seen, Theorem 2.1 states that if E has the same perimeter of its Steiner symmetral E^s, then almost every section of E in the y direction is a segment and the two normals at the endpoints of the segment are symmetric. However, this is not enough to conclude that E coincides with E^s (up to a transaltion), as it is clear by looking at the picture below.

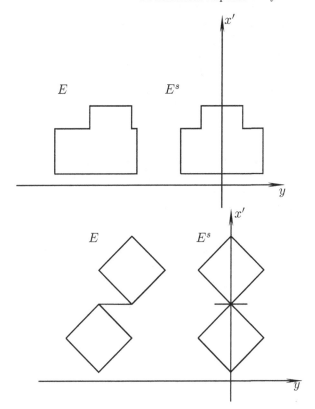

Thus, in order to deduce from the equality $P(E) = P(E^s)$ that E and E^s coincide, up to a translation in the y direction, we need to make some assumption on the set E or on E^s. To this aim let us start by assuming that, given an open set $U \subset \mathbb{R}^{n-1}$,

$$(H_1) \qquad \mathcal{H}^{n-1}(\{x \in \partial^* E^s : \nu_n^{E^s}(x) = 0\} \cap (U \times \mathbb{R})) = 0,$$

i.e., the (reduced) boundary of E^s has no flat parts parallel to the y direction. Notice that this assumption rules out the example shown on the upper part of the picture. Moreover, as we shall see in a moment, (H_1) holds in an open set U if and only if the distribution function is a $W^{1,1}$ function in U. To this aim, let us recall the following well known result concerning the graph of a BV function (see, for instance, [18, Ch. 4, Sec. 1.5, Th. 1, and Ch. 4, Sec. 2.4, Th. 4]).

Theorem 2.2. *Let $U \subset \mathbb{R}^{n-1}$ be a bounded open set and $u \in L^1(U)$. Then the subgraph of U,*

$$\mathcal{S}_u = \{(x', y) \in U \times \mathbb{R} : y < u(x')\},$$

is a set of finite perimeter in $U \times \mathbb{R}$ if and only if $u \in BV(U)$. Moreover, in this case,

$$P(\mathcal{S}_u; B \times \mathbb{R}) = \int_B \sqrt{1 + |\nabla u|^2} dx' + |D^s u|(B) \qquad (2.6)$$

for every Borel set $B \subset U$.

Notice that if E is a bounded set of finite perimeter, since $\mu \in BV(\mathbb{R}^{n-1})$ by Lemma 1.1, and

$$E^s = \{(x', y) \in \mathbb{R}^{n-1} \times \mathbb{R} : -\mu(x')/2 < y < \mu(x')/2\}, \qquad (2.7)$$

from Theorem 2.2 we get immediately that E^s is a set of finite perimeter, being the intersection of the two sets of finite perimeter $\{(x', y) : y > -\mu(x')/2\}$ and $\{(x', y) : y < \mu(x')/2\}$.

Proposition 2.1. *Let E be any set of finite perimeter in $\mathbb{R}^n$, $n \geq 2$, with finite measure. Let U be an open subset of $\mathbb{R}^{n-1}$. Then the following conditions are equivalent:*

(i) $\mathcal{H}^{n-1}(\{x \in \partial^* E^s : \nu_n^{E^s}(x) = 0\} \cap (U \times \mathbb{R})) = 0$,
(ii) $P(E^s; B \times \mathbb{R}) = 0$ for every Borel set $B \subset U$ such that $\mathcal{L}^{n-1}(B) = 0$,
(iii) $\mu \in W^{1,1}(U)$.

Proof. Let us assume that (i) holds and fix a Borel set $B \subset U$ such that $\mathcal{L}^{n-1}(B) = 0$. Using coarea formula (1.10) we get

$$\begin{aligned}
P(E^s; B \times \mathbb{R}) &= \mathcal{H}^{n-1}(\partial^* E^s \cap (B \times \mathbb{R})) \\
&= \mathcal{H}^{n-1}(\{x \in \partial^* E^s : \nu_n^{E^s}(x) \neq 0\} \cap (B \times \mathbb{R})) \\
&= \int_{\partial^* E^s} \frac{1}{|\nu_n^{E^s}(x)|} \chi_{\{\nu_n^{E^s} \neq 0\} \cap (B \times \mathbb{R})}(x) |\nu_n^{E^s}(x)| \, d\mathcal{H}^{n-1} \\
&= \int_B dx' \int_{(\partial^* E^s)_{x'}} \frac{\chi_{\{\nu_n^{E^s} \neq 0\}}(x', y)}{|\nu_n^{E^s}(x', y)|} \, d\mathcal{H}^0(y) = 0,
\end{aligned}$$

hence (ii) follows.

If (ii) holds and B is a null set in U, by applying (1.14) with E replaced by E^s we get $|D\mu|(B) = 0$. Thus, $D\mu$ is absolutely continuous with respect to $\mathcal{L}^{n-1}$, hence $\mu \in W^{1,1}(U)$.

Notice that, if E_1, E_2 are two sets of finite perimeter and B is an open set, then (see [1, Proposition 3.38]) $P(E_1 \cap E_2; B) \leq P(E_1; B) + P(E_2; B)$ and, by approximation, the same inequality holds also when B is a Borel set. Therefore, recalling (2.7) and (2.6) we get that, if (iii) holds, for any Borel set B in U

$$P(E^s; B \times \mathbb{R}) \leq 2P(\mathcal{S}_{\mu/2}; B \times \mathbb{R}) = \int_B \sqrt{4 + |\nabla \mu|^2} \, dx'. \qquad (2.8)$$

Set $B_0 = \pi(\partial^* E^s) \setminus G_{E^s}$, where $G_{E^s} \subset \pi(E)^+$ is a Borel set satisfying (1.20) with E replaced by E^s. Since by Theorem 1.3 $(\partial^* E^s)_{x'} = \emptyset$ for $\mathcal{L}^{n-1}$-a.e.

$x' \notin \pi(E)^+$, we have $\mathcal{L}^{n-1}(B_0) = \mathcal{L}^{n-1}(\pi(\partial^* E^s) \setminus \pi(E)^+) + \mathcal{L}^{n-1}(\pi(E)^+ \setminus G_{E^s}) = 0$. Therefore, from (2.8) we get that $P(E^s; (B_0 \cap U) \times \mathbb{R}) = 0$, i.e. $\mathcal{H}^{n-1}((\partial^* E^s \setminus (G_{E^s} \times \mathbb{R})) \cap (U \times \mathbb{R})) = 0$. Then, (i) follows since by definition $\{x \in \partial^* E^s : \nu_n^{E^s}(x) = 0\} \subset \partial^* E^s \setminus (G_{E^s} \times \mathbb{R})$. $\square$

It may seem strange that assumption (H_1) is made on the Steiner symmetral E^s. Alternatively, we could make a similar assumption on E by requiring that

$$(H_1') \qquad \mathcal{H}^{n-1}(\{x \in \partial^* E : \nu_n^E(x) = 0\} \cap (U \times \mathbb{R})) = 0.$$

Actually, it is not difficult to show that (H_1') implies (H_1), while the converse is false in general, as one can see by simple examples. In fact, if (H_1') holds, arguing exactly as in the proof of the implication '(i)$\Rightarrow$(ii)' in Proposition 2.1 we get that $P(E; B \times \mathbb{R}) = 0$ for any Borel set $B \subset U$ with zero measure. Then (2.1) implies that the same property holds also for E^s and thus, by Proposition 2.1, we get that E^s satisfies (H_1). Notice also that when $P(E) = P(E^s)$, then by (2.1) we have that $P(E; B \times \mathbb{R}) = P(E^s; B \times \mathbb{R})$ for any Borel set $B \subset \mathbb{R}^{n-1}$. Thus one immediately gets that in this case the two conditions (H_1), (H_1') are equivalent.

Let us now comment on the example on the lower part of the picture above. It is clear that in that case things go wrong, in the sense that E and E^s are not equal, because even though the set E is connected in a strict topological sense it is 'essentially disconnected'. Therefore, to deal with similar examples one could device to use a suitable notion of connectedness set up in the context of sets of finite perimeter (see, for instance, [1, Example 4.18]). However, we will not follow this path. Instead, we will use the information provided by Proposition 2.1.

If the distribution function μ is of class $W^{1,1}(U)$, then for $\mathcal{H}^{n-2}$-a.e. $x' \in U$ we can define its *precise representative* $\widetilde{\mu}(x')$ (see [15] or [27]) as the unique value such that

$$\lim_{r \to 0} \fint_{B_r^{n-1}(x')} |\mu(y) - \widetilde{\mu}(x')| \, dx' = 0, \qquad (2.9)$$

where by $B_r^{n-1}(x')$ we have denoted the $(n-1)$-dimensional ball with centre x' and radius r. Then, in order to rule out a situation like the one on the bottom of the picture above, we make the assumption

$$(H_2) \qquad \widetilde{\mu}(x') > 0 \qquad \text{for } \mathcal{H}^{n-2}\text{-a.e. } x' \in U.$$

Next result, proved in [9], shows that the two examples in the picture are indeed the only cases where the equality $P(E) = P(E^s)$ does not imply that the two sets are equal. As for Theorem 1.1, we state the result in a local form.

Theorem 2.3. *Let E be a set of finite perimeter $\mathbb{R}^n$, with $n \geq 2$, such that*

$$P(E^s) = P(E). \qquad (2.10)$$

Let us assume that (H_1) and (H_2) hold in some open set $U \subset \mathbb{R}^{n-1}$. Then, for every connected open subset U_α of U, $E \cap (U_\alpha \times \mathbb{R})$ is equivalent to $E^s \cap (U_\alpha \times \mathbb{R})$, up to a translation in the y direction.

In particular, if (H_1) and (H_2) hold in a connected open set U such that $\mathcal{L}^{n-1}(\pi(E)^+ \setminus U) = 0$, then E is equivalent to E^s, up to a translation in the y direction.

As far as I know, this result was known in the literature only for convex sets, where it can be proved with a simple argument. In fact, let us assume that E is an open convex set such that $P(E) = P(E^s) < \infty$. Then, $\pi(E)$ is also an open convex set and there exist two functions $y_1, y_2 : \pi(E) \to \mathbb{R}$, y_1 convex, and y_2 concave, such that

$$E = \{(x', y) : x' \in \pi(E), y_1(x') < y < y_2(x')\}.$$

Let us now fix an open set $U \subset\subset \pi(E)$. From assumption (2.10) and from (2.1) we have that $P(E^s; U \times \mathbb{R}) = P(E; U \times \mathbb{R})$. Since y_1 and y_2 are Lipschitz continuous in U, we can write this equality as

$$2 \int_U \sqrt{1 + \frac{|\nabla(y_2 - y_1)|^2}{4}} \, dx' = \int_U \sqrt{1 + |\nabla y_1|^2} \, dx' + \int_U \sqrt{1 + |\nabla y_1|^2} \, dx'.$$

From this equality, the strict convexity of the function $t \mapsto \sqrt{1 + t^2}$ and the arbitrariness of U, we get that $\nabla y_2 = -\nabla y_1$ in $\pi(E)$ and thus $y_2 = -y_1 + const.$. This shows that E coincides with E^s, up to a translation in y direction.

The proof of Theorem 2.3, for which we refer to [9], uses delicate tools from Geometric Measure theory. However, in the special case considered below it can be greatly simplified.

Proof of Theorem 2.3 in a Special Case. Let us assume that E is an open set, that $\pi(E)$ is connected and that E is bounded in y direction. Notice that since E is open, then μ is a lower semicontinuous function and for any open set $U \subset\subset \pi(E)$ there exists a constant $c(U) > 0$ such that

$$\mu(x') \geq c(U) \qquad \text{for all } x' \in U. \tag{2.11}$$

Moreover, since E is bounded in the y direction, the function

$$x' \in \pi(E) \mapsto m(x') = \int_{E_{x'}} y \, dy$$

is bounded in $\pi(E)$. Then, the same arguments used in the proofs of Lemmas 1.1 and 1.3 yield that $m \in BV_{\text{loc}}(\pi(E))$ and that for $\mathcal{L}^{n-1}$-a.e. $x' \in \pi(E)$, $i = 1, \ldots, n-1$,

$$\frac{\partial m}{\partial x_i}(x') = \sum_{y \in \partial^* E_{x'}} \frac{y \, \nu_i^E(x', y)}{|\nu_n^E(x', y)|}, \tag{2.12}$$

where we have denoted by $\partial m/\partial x_i$ the absolutely continuous part of the derivative $D_i m$.

By Proposition 2.1 we have that (H_1) implies that the distribution function μ is a Sobolev function. The same assumption implies also that $m \in W^{1,1}_{\text{loc}}(\pi(E))$. In fact, the argument used to prove (1.14) shows that if $B \subset \pi(E)$ is a Borel set, then $|Dm|(B) \leq MP(E; B \times \mathbb{R})$, where M is a constant such that $E \subset \mathbb{R}^{n-1} \times (-M, M)$. Therefore, if $\mathcal{L}^{n-1}(B) = 0$, from (2.1) and Proposition 2.1 we have that $P(E; B \times \mathbb{R}) = P(E^s; B \times \mathbb{R}) = 0$, hence Dm is absolutely continuous with respect to $\mathcal{L}^{n-1}$.

Let us now denote, for any $x' \in \pi(E)$ by $b(x')$ the baricenter of the section $E_{x'}$, i.e.,

$$b(x') = \frac{\displaystyle\int_{E_{x'}} y\,dy}{\mu(x')}\,.$$

From (2.11) and Proposition 2.1 we have that b too belongs to the space $W^{1,1}_{\text{loc}}(\pi(E))$. Thus, to prove the assertion, since $\pi(E)$ is a connected open set, it is enough to show that $\nabla b \equiv 0$, hence b is constant on $\pi(E)$. To this aim, let us evaluate the partial derivatives of b, using the representation formulas (2.12) and (1.21). We have, for any $i = 1, \ldots, n-1$ and for $\mathcal{L}^{n-1}$-a.e. $x' \in \pi(E)$,

$$\frac{\partial b}{\partial x_i}(x') = \frac{1}{\mu(x')}\left(\sum_{y \in \partial^* E_{x'}} \frac{y\,\nu_i^E(x', y)}{|\nu_n^E(x', y)|} - \frac{\displaystyle\int_{E_{x'}} y\,dy}{\mu(x')}\sum_{y \in \partial^* E_{x'}}\frac{\nu_i^E(x', y)}{|\nu_n^E(x', y)|}\right). \quad (2.13)$$

Since for $\mathcal{L}^{n-1}$-a.e. $x' \in \pi(E)$ (2.4) and (2.5) hold, the right hand side of (2.13) is equal to

$$\frac{1}{\mu(x')}\left[\left(y_2(x') + y_1(x')\right)\frac{\nu_i^E(x', y_2(x'))}{|\nu_n^E(x', y_2(x'))|} - \frac{1}{2}\frac{y_2^2(x') - y_1^2(x')}{y_2(x') - y_1(x')}\frac{2\nu_i^E(x', y_2(x'))}{|\nu_n^E(x', y_2(x'))|}\right] = 0.$$

Hence, the assertion follows. $\square$

3 The Pòlya–Szegö Inequality

We are going to present the classical Pòlya–Szegö inequality for the spherical rearrangement of a Sobolev function u and discuss what can be said about the function u when the equality holds. In order to simplify the exposition, we shall assume that u is a nonnegative measurable function from $\mathbb{R}^n$, with compact support. However, most of the results presented here, like Theorem 3.1, still hold with no restrictions on the support or on the sign of u.

Given u, we set, for any $t \geq 0$,

$$\mu_u(t) = \mathcal{L}^n(\{x \in \mathbb{R}^n : u(x) > t\}).$$

The function μ_u is called the *distribution function* of u. Clearly μ_u is a decreasing, right-continuous function such that

$$\mu_u(0) = \mathcal{L}^n(\text{supp}\,u), \quad \mu_u(\text{esssup}\,u) = 0, \quad \mu_u(t-) = \mathcal{L}^n(\{u \geq t\}) \quad \text{for all } t > 0.$$
(3.1)

Notice that from the last equality we have that when $t > 0$

$$\mu_u \text{ is continuous in } t \text{ iff } \mathcal{L}^n(\{u = t\}) = 0.$$

Let us now introduce the *decreasing rearrangement* of u, that is the function $u^* : [0, +\infty) \rightarrow [0, +\infty)$ defined, for any $s \geq 0$, by setting

$$u^*(s) = \sup\{t \geq 0 : \mu_u(t) > s\}.$$

Clearly, u^* is a decreasing, right-continuous function. The following elementary properties of u^* are easily checked:

(j) $u^*(\mu_u(t)) \leq t \leq u^*(\mu_u(t)-)$ for all $0 \leq t < \text{esssup}\,u$;
(jj) $\mu_u(u^*(s)) \leq s \leq \mu_u(u^*(s)-)$ for all $0 \leq s < \mathcal{L}^n(\text{supp}\,u)$;
(jjj) $\mathcal{L}^1(\{s : u^*(s) > t\} = \mu_u(t)$ for all $t \geq 0$.

Notice that (jjj) states that the functions u and u^* are *equi-distributed*, i.e., $\mu_u = \mu_{u^*}$. Let us now define the *spherical symmetric rearrangement* of u, that is the function $u^\star : \mathbb{R}^n \rightarrow [0, +\infty)$, such that for all $x \in \mathbb{R}^n$

$$u^\star(x) = u^*(\omega_n |x|^n).$$
(3.2)

By definition and by (jjj) we have

$$\mathcal{L}^n(\{u^\star > t\}) = \mathcal{L}^n(\{u > t\}) \qquad \text{for all } t \geq 0,$$

i.e., $\mu_u = \mu_{u^\star}$. Thus also u and $u^\star$ are equi-distributed. As a simple consequence of this equality and Fubini's theorem we have, for all $p \geq 1$,

$$\int_{\mathbb{R}^n} |u^\star(x)|^p \, dx = \int_{\mathbb{R}^n} |u(x)|^p \, dx,$$

and, letting $p \rightarrow +\infty$, $\text{esssup}\,u = \text{esssup}\,u^\star$.

If u is a smooth function, in general its symmetric rearrangement will be no longer smooth (actually the best we may expect from Theorem 3.1 below is that $u^\star$ is Lipschitz continuous). However, the symmetric rearrangement behaves nicely on Sobolev functions, as shown by the next result.

Theorem 3.1 (Pòlya–Szegö Inequality). *Let* $u \in W^{1,p}(\mathbb{R}^n)$, $p \geq 1$, *be a nonnegative function with compact support. Then* $u^\star \in W^{1,p}(\mathbb{R}^n)$ *and*

$$\int_{\mathbb{R}^n} |\nabla u^\star(x)|^p \, dx \leq \int_{\mathbb{R}^n} |\nabla u(x)|^p \, dx.$$
(3.3)

The proof of this result relies upon two main ingredients, the isoperimetric inequality and the coarea formula for BV functions. Let us start by recalling the latter.

Let u be a $BV(\Omega)$ function. Then, for $\mathcal{L}^1$-a.e. $t \in \mathbb{R}$, the set $\{u > t\}$ has finite perimeter in Ω. Moreover, for any Borel function $g : \Omega \to [0, +\infty]$, the following formula holds (see [1, Theorem 3.40]).

$$\int_\Omega g(x)\, d|Du| = \int_{-\infty}^{\infty} dt \int_{\partial^* \{u > t\}} g(x)\, d\mathcal{H}^{n-1}. \tag{3.4}$$

In the special case $u \in W^{1,1}(\Omega)$, it can be shown that for $\mathcal{L}^1$-a.e. t the reduced boundary $\partial^* \{u > t\}$ coincides, modulo a set of $\mathcal{H}^{n-1}$-measure zero, with the level set $\{\tilde{u} = t\}$, where $\tilde{u}$ denotes the *precise representative* of u, which is defined $\mathcal{H}^{n-1}$-a.e. in Ω as in (2.9). Therefore, if $u \in W^{1,1}(\Omega)$, (3.4) becomes

$$\int_\Omega g(x)|\nabla u(x)|\, dx = \int_{-\infty}^{\infty} dt \int_{\{\tilde{u}=t\}} g(x)\, d\mathcal{H}^{n-1}(x). \tag{3.5}$$

The *isoperimetric inequality* states that if E be a set of finite perimeter, then

$$\left(\min\{\mathcal{L}^n(E), \mathcal{L}^n(\mathbb{R}^n \setminus E)\}\right)^{\frac{n-1}{n}} \leq \frac{1}{n\omega_n^{1/n}} P(E). \tag{3.6}$$

Moreover, the equality holds if and only if E is (equivalent to) a ball.

Next lemma shows that if u is a Sobolev function, then the same is also true for $u^\star$.

Lemma 3.1. *Let u be a nonnegative function with compact support from the space $W^{1,1}(\mathbb{R}^n)$. Then $u^\star$ belongs to $W^{1,1}(\mathbb{R}^n)$ and*

$$\int_{\mathbb{R}^n} |\nabla u^\star|\, dx \leq \int_{\mathbb{R}^n} |\nabla u|\, dx. \tag{3.7}$$

Proof. Let us first prove that for any $0 < a < b$, the function u^* is absolutely continuous in (a, b) and

$$\int_a^b |u^{*\prime}(s)|\, ds \leq \frac{1}{n\omega_n^{1/n} a^{\frac{n-1}{n}}} \int_{\{u^*(b) < u < u^*(a)\}} |\nabla u|\, dx. \tag{3.8}$$

To this aim, we start by observing that from the third equality in (3.1) and from inequality (jj) we have

$$\mathcal{L}^n(\{x \in \mathbb{R}^n : u^*(b) < u(x) < u^*(a)\}) = \mu_u(u^*(b)) - \mu_u(u^*(a)-) \leq b - a. \tag{3.9}$$

Let us denote by $\omega : [0, \infty) \to [0, +\infty)$ the modulus of continuity of the integral of $|\nabla u|$, i.e., a continuous function, vanishing at zero and such that for any set of finite measure E

$$\int_E |\nabla u|\,dx \le \omega(\mathcal{L}^n(E)).$$

Using the coarea formula (3.5), the isoperimetric inequality (3.6) and (jj) again, we obtain the following estimate for the integral of $|\nabla u|$ between two level sets,

$$\int_{\{u^*(b)<u<u^*(a)\}} |\nabla u|\,dx = \int_{u^*(b)}^{u^*(a)} P(\{u>t\})\,dt$$

$$\ge n\omega_n^{1/n} \int_{u^*(b)}^{u^*(a)} \left(\mathcal{L}^n(\{u>t\})\right)^{\frac{n-1}{n}} dt$$

$$\ge n\omega_n^{1/n}[\mu_u(u^*(a)-)]^{\frac{n-1}{n}}(u^*(a)-u^*(b)) \quad (3.10)$$

$$\ge n\omega_n^{1/n}a^{\frac{n-1}{n}}(u^*(a)-u^*(b)).$$

Let us now take a finite number of pairwise disjoint intervals $(a_i,b_i) \subset (a,b)$, $i=1,\ldots,N$. By applying (3.9) and (3.10) to each interval (a_i,b_i), we get

$$\sum_{i=1}^N |u^*(b_i)-u^*(a_i)| \le \frac{1}{n\omega_n^{1/n}a^{\frac{n-1}{n}}} \sum_{i=1}^N \int_{\{u^*(b_i)<u<u^*(a_i)\}} |\nabla u|\,dx$$

$$\le \frac{1}{n\omega_n^{1/n}a^{\frac{n-1}{n}}} \omega\left(\sum_{i=1}^N (b_i-a_i)\right). \quad (3.11)$$

From this inequality it follows immediately that u is absolutely continuous in (a,b), since the left hand side is smaller than a given $\varepsilon > 0$ as soon as the sum of the lengths of the intervals (a_i,b_i) is sufficiently small. Moreover, by taking the supremum of the left hand side of (3.11) over all possible partitions of the interval (a,b), from the first inequality in (3.11) we get immediately (3.8). Notice that from (3.8) it follows that u^* is in $W^{1,1}_{loc}(\mathbb{R}^n \setminus \{0\})$. To prove the assertion, we fix $\sigma > 1$ and estimate the integral of $|\nabla u|$ in the annuli $A_{k,\sigma} = \{x \in \mathbb{R}^n : \omega_n^{-1/n}\sigma^{k/n} < |x| < \omega_n^{-1/n}\sigma^{(k+1)/n}\}$, for $k \in \mathbb{Z}$. Using (3.8) again, and recalling the definition (3.2), we get, for any $k \in \mathbb{Z}$,

$$\int_{A_{k,\sigma}} |\nabla u^*|\,dx = n\omega_n \int_{A_{k,\sigma}} |x|^{n-1}|u^{*\prime}(\omega_n|x|^n)|\,dx$$

$$= n^2\omega_n^2 \int_{\omega_n^{-1/n}\sigma^{k/n}}^{\omega_n^{-1/n}\sigma^{(k+1)/n}} r^{2n-2}|u^{*\prime}(\omega_n r^n)|\,dr$$

$$= n\omega_n^{1/n} \int_{\sigma^k}^{\sigma^{k+1}} s^{\frac{n-1}{n}}|u^{*\prime}(s)|\,ds$$

$$\le \sigma^{\frac{n-1}{n}} \int_{\{u^*(\sigma^k)<u<u^*(\sigma^{k+1})\}} |\nabla u|\,dx.$$

Then the assertion immediately follows by summing up both sides of this inequality over all $k \in \mathbb{Z}$ and then letting $\sigma \to 1^+$. $\square$

Notice that the lemma we have just proved provides the Pólya–Szegö inequality for $p = 1$. However, for the general case $p \geq 1$ we present a different proof which has the advantage of giving better information when the inequality becomes an equality.

To this aim let us introduce a few quantities that will be useful later. If $u \in W^{1,1}_{\mathrm{loc}}(\mathbb{R}^n)$, we set

$$\mathcal{D}^+_u = \{x \in \mathbb{R}^n : \nabla u(x) \neq 0\}, \qquad \mathcal{D}^0_u = \mathbb{R}^n \setminus \mathcal{D}^+_u .$$

We can now give a representation formula for the derivative of μ_u. Notice that the formula stated in (3.12) uses the fact that the $|\nabla u^\star|$ is $\mathcal{H}^{n-1}$-a.e. constant on $\mathcal{L}^1$-a.e. level set $\{u^\star = t\}$.

Lemma 3.2. Let $u \in W^{1,1}(\mathbb{R}^n)$ be a nonnegative function with compact support. Then, for $\mathcal{L}^1$-a.e. $t > 0$,

$$\mu'_u(t) = -\frac{\mathcal{H}^{n-1}(\{u^\star = t\})}{|\nabla u^\star|_{|\{u^\star = t\}}} \leq -\int_{\{\tilde{u}=t\}} \frac{1}{|\nabla u|}\, d\mathcal{H}^{n-1} . \qquad (3.12)$$

Proof. First of all let us evaluate $\mu_u(t)$ using the coarea formula (3.5). We get, for all $t \geq 0$,

$$\mu_u(t) = \mathcal{L}^n\left(\{u > t\} \cap \mathcal{D}^0_u\right) + \mathcal{L}^n\left(\{u > t\} \cap \mathcal{D}^+_u\right)$$

$$= \mathcal{L}^n\left(\{u > t\} \cap \mathcal{D}^0_u\right) + \int_{\mathcal{D}^+_u} \chi_{\{u>t\}}(x)\, dx \qquad (3.13)$$

$$= \mathcal{L}^n\left(\{u > t\} \cap \mathcal{D}^0_u\right) + \int_t^{+\infty} ds \int_{\{\tilde{u}=s\}} \frac{\chi_{\mathcal{D}^+_u}}{|\nabla u|}\, d\mathcal{H}^{n-1}$$

$$= \mathcal{L}^n\left(\{u > t\} \cap \mathcal{D}^0_u\right) + \int_t^{+\infty} ds \int_{\{\tilde{u}=s\}} \frac{1}{|\nabla u|}\, d\mathcal{H}^{n-1} ,$$

where the last equality follows by observing that coarea formula (3.5) implies that $\mathcal{H}^{n-1}(\{\tilde{u} = t\} \cap \mathcal{D}^0_u) = 0$ for $\mathcal{L}^1$-a.e. $t \geq 0$. By applying (3.13) to $u^\star$, we get also that for all $t \geq 0$

$$\mu_u(t) = \mathcal{L}^n\left(\{u^\star > t\} \cap \mathcal{D}^0_{u^\star}\right) + \int_t^{+\infty} \frac{\mathcal{H}^{n-1}(\{u^\star = t\})}{|\nabla u^\star|_{|\{u^\star=t\}}} \qquad (3.14)$$

Let us now recall a nice property of absolutely continuous functions (see for instance [10, Lemma 2.4]).

If g is an absolutely continuous function in a bounded open interval I and, for all $t \in \mathbb{R}$, we set $\phi_g(t) = \mathcal{L}^1(\{g > t\} \cap \mathcal{D}^0_g)$, then ϕ_g is a nondecreasing function such that $\phi'_g(t) = 0$ for $\mathcal{L}^1$-a.e. t.

By applying this result with $g = u^\star$ and observing that $\mathcal{L}^n\left(\{u^\star > t\} \cap \mathcal{D}^0_{u^\star}\right) = \mathcal{L}^1\left(\{u^\star > t\} \cap \mathcal{D}^0_{u^\star}\right)$, for all $t > 0$, from (3.14) we get immediately the equality in (3.12). On the other hand, the inequality

$$\mu_{\tilde{u}}'(t) \leq -\int_{\{\tilde{u}=t\}} \frac{1}{|\nabla u|} \, d\mathcal{H}^{n-1}$$

follows immediately from (3.13). □

We are now ready to prove the Pólya–Szegő inequality (3.3).

Proof of Theorem 3.1. Let us fix a nonnegative function $u \in W^{1,p}(\mathbb{R}^n)$ with compact support and let us assume, without loss of generality, that u coincides with its precise representative $\tilde{u}$. From Lemma 3.1 we know already that $u^\star$ belongs to the space $W^{1,1}(\mathbb{R}^n)$. Thus, using the coarea formula (3.5) with u replaced by $u^\star$ and recalling that $|\nabla u^\star|$ is constant on the level sets of $u^\star$, we get

$$\int_{\mathbb{R}^n} |\nabla u^\star|^p \, dx = \int_0^{+\infty} dt \int_{\{u^\star=t\}} |\nabla u^\star|^{p-1} \, d\mathcal{H}^{n-1}$$

$$= \int_0^{+\infty} \mathcal{H}^{n-1}(\{u^\star = t\}) |\nabla u^\star|^{p-1}_{|\{u^\star=t\}} \, dt \, .$$

From this equation, using twice (3.12), the isoperimetric inequality (3.6), Hölder's inequality and coarea formula again, we get

$$\int_{\mathbb{R}^n} |\nabla u^\star|^p \, dx = \int_0^{+\infty} \frac{\left[\mathcal{H}^{n-1}(\{u^\star = t\})\right]^p}{[-\mu_u'(t)]^{p-1}} \, dt$$

$$\leq \int_0^{+\infty} \frac{\left[\mathcal{H}^{n-1}(\{u^\star = t\})\right]^p}{\left(\int_{\{u^\star=t\}} \frac{d\mathcal{H}^{n-1}}{|\nabla u|}\right)^{p-1}} \, dt \qquad (3.15)$$

$$\leq \int_0^{+\infty} \frac{\left[\mathcal{H}^{n-1}(\{u = t\})\right]^p}{\left(\int_{\{u=t\}} \frac{d\mathcal{H}^{n-1}}{|\nabla u|}\right)^{p-1}} \, dt$$

$$\leq \int_0^{+\infty} dt \int_{\{u=t\}} |\nabla u|^{p-1} \, d\mathcal{H}^{n-1} = \int_{\mathbb{R}^n} |\nabla u|^p \, dx \, .$$

Hence (3.3) follows. □

Let us now discuss the equality case in (3.3). First, notice that if this is the case, then all inequalities in (3.15) are in fact equalities. In particular, if the second inequality in (3.15) holds as an equality, then we can conclude that the set $\{u > t\}$ is (equivalent to) a ball for $\mathcal{L}^1$-a.e. $t \geq 0$. Moreover, if the equality holds in the third inequality (where we have used Hölder inequality), the conclusion is that for $\mathcal{L}^1$-a.e. $t \geq 0$, $|\nabla u|$ is $\mathcal{H}^{n-1}$-a.e. constant on the level set $\{u = t\}$.

These are the immediate consequences of the equality case. However, with some extra work, one can prove the following, more precise, result (see [5] or [11, Theorem 2.3]).

Proposition 3.1. *Let* $u \in W^{1,p}(\mathbb{R}^n)$, $p \geq 1$, *a nonnegative function with compact support such that*

$$\int_{\mathbb{R}^n} |\nabla u^\star|^p \, dx = \int_{\mathbb{R}^n} |\nabla u|^p \, dx. \tag{3.16}$$

Then there exist a function v, *equivalent to* u, *i.e. such that* $v(x) = u(x)$ *for* $\mathcal{L}^n$-*a.e.* $x \in \mathbb{R}^n$, *and a family of open balls* $\{U_t\}_{t \geq 0}$ *such that:*

(i) $\{v > t\} = U_t$ *for* $t \in [0, \operatorname{esssup} u)$;

(ii) $\{v = \operatorname{ess\,sup} u\} = \displaystyle\bigcap_{0 \leq t < \operatorname{esssup} u} \overline{U}_t$,   *and is a closed ball (possibly a point);*

(iii) v *is lower semicontinuous in* $\{v < \operatorname{esssup} u\}$;

(iv)   *if* $v(x) \in (0, \operatorname{esssup} u)$ *and* $\mathcal{L}^n(\{u = v(x)\}) = 0$, *then* $x \in \partial U_{v(x)}$;

(v)   *for every* $t \in (0, \operatorname{esssup} u)$ *there exists at most one point* $x \in \partial U_t$ *such that* $v(x) \neq t$;

(vi)   *the coarea formula (3.5) holds with* $\widetilde{u}$ *replaced by* v;

(vii)   *for* $\mathcal{L}^1$-*a.e.* $t \in (0, \operatorname{esssup} u)$, $|\nabla v(x)| = |\nabla u^\star||_{\{u^\star = t\}}$ *for* $\mathcal{H}^{n-1}$-*a.e.* $x \in \partial U_t$.

This proposition contains all the information that we can extract from equality (3.16). However it is not true in general that (3.16) implies that u coincides with $u^\star$, up to a translation in x. This can be easily seen by considering any spherically symmetric nonnegative function w, such that $\mathcal{L}^n(\{w = t_0\}) > 0$ for some $t_0 \in (0, \operatorname{esssup} w)$ and another function u whose graph agrees with the graph of w where $u < t_0$ and with a slight translated of the graph of w where $u > t_0$. Then $u^\star = w$ and (3.16) holds, but u is not spherically symmetric. What goes wrong in this example is the fact that the gradient of w (and of u) vanishes in a set of positive measure. Thus, this example suggests to introduce the following assumption,

(H) $\mathcal{L}^n(\{0 < u^\star < \operatorname{esssup} u\} \cap \mathcal{D}^0_{u^\star}) = 0$.

Notice that we are in a situation similar to the one we were in the previous lecture when dealing with the assumption (H_1). In fact, it can be proved (see for instance [10, Lemma 3.3]) that (H) is implied by the stronger assumption

(H') $\mathcal{L}^n(\{0 < u < \operatorname{esssup} u\} \cap \mathcal{D}^0_u) = 0$.

Moreover, (H) is equivalent to the absolute continuity in $(0, +\infty)$ of the distribution function μ_u and the two conditions (H) and (H') are equivalent if (3.16) holds (see [10, Lemma 3.3] again).

The following result was proved for the first time in the Sobolev setting by Brothers and Ziemer ([5]). It shows that when equality holds in (3.3), assumption (H) guarantees that u and $u^\star$ agree.

Theorem 3.2. *Let $u \in W^{1,p}(\mathbb{R}^n)$, $p > 1$, be a nonnegative function with compact support such that (3.16) and (H) hold. Then $u^* = u$, up to a translation in x.*

Notice that the above result is in general false if $p = 1$, even in one dimension. To see this it is enough to take a function which is increasing in the interval $(-\infty, a)$ and decreasing in $(a, +\infty)$.

The starting point of the proof of Brothers and Ziemer is to observe, as we have done before, that the equality (3.1) yields that $\mathcal{L}^n$-a.e. set $\{u > t\}$ is a ball and that $|\nabla u|$ is constant on the corresponding boundary. Then the difficult part of the proof consists in exploiting assumption (H) to deduce that all these balls are concentric, i.e. u is spherically symmetric. To prove this, we shall not follow the original argument contained in ([5]), but a somewhat simpler one used in [11], which is in turn inspired to an alternative proof of Theorem 3.2 given in [17].

To this aim from now on we shall assume, without loss of generality, that u agrees with the representative v provided by Proposition 3.1 and that $U_0 = \{u > 0\}$ is a ball centered at the origin. Then, for all $0 < t < \operatorname{esssup} u$ we denote by R_t the radius of the ball U_t and set, for all $x \in U_0$,

$$\Phi(x) = \left(\frac{\mu_u(u(x))}{\omega_n} \right)^{1/n} . \tag{3.17}$$

To understand the role of the function Φ, observe that if $x \in U_0$ is a point such that $u(x) = t$, then $\mu_u(u(x)) = \mathcal{L}^n(U_t)$. Therefore, $\Phi(x)$ is equal to the radius of the ball U_t.

The following lemma is a crucial step toward the proof of Theorem 3.2.

Lemma 3.3. *Under the assumptions of Theorem 3.2, $\Phi \in W^{1,\infty}(U_0)$ and*

$$|\nabla \Phi(x)| = 1 \qquad \text{for } \mathcal{L}^n\text{-a.e. } x \in U_0 \setminus \{u = \operatorname{esssup} u\} . \tag{3.18}$$

Proof. We claim that $\mu_u \circ u \in W^{1,\infty}(U_0)$ and that

$$\nabla(\mu_u \circ u)(x) = -\mathcal{H}^{n-1}(\{u = u(x)\}) \frac{\nabla u(x)}{|\nabla u|_{|\{u=u(x)\}}} \chi_{D_u^+}(x) \qquad \text{for } \mathcal{L}^n\text{-a.e. } x \in U_0. \tag{3.19}$$

From assumption (H), which is equivalent by (3.16) to (H'), using (3.13) and Proposition 3.1 (vii), we get that for all $0 \leq t \leq \operatorname{esssup} u$,

$$\mu_u(t) = \mathcal{L}^n(\{u = \operatorname{esssup} u\}) + \int_t^{+\infty} \frac{\mathcal{H}^{n-1}(\{u = s\})}{|\nabla u|_{|\{u=s\}}} \, ds .$$

For $\varepsilon > 0$, we set

$$\mu_{u,\varepsilon}(t) = \mathcal{L}^n(\{u = \operatorname{esssup} u\}) + \int_t^{+\infty} \frac{\mathcal{H}^{n-1}(\{u = s\})}{|\nabla u|_{|\{u=s\}} + \varepsilon} \, ds .$$

Clearly, $\mu_{u,\varepsilon}(t) \uparrow \mu_u(t)$ for every $t \geq 0$ as $\varepsilon \downarrow 0$. Moreover, $\mu_{u,\varepsilon}$ is Lipschitz continuous in $[0, +\infty)$, and

$$\mu'_{u,\varepsilon}(t) = -\frac{\mathcal{H}^{n-1}(\{u=t\})}{|\nabla u|_{|\{u=t\}} + \varepsilon} \qquad \text{for } \mathcal{L}^1\text{-a.e. } t \geq 0,$$

whence

$$|\mu'_{u,\varepsilon}(t) - \mu'_u(t)| \leq \frac{\varepsilon \mathcal{H}^{n-1}(\{u=t\})}{|\nabla u|_{|\{u=t\}} \left(|\nabla u|_{|\{u=t\}} + \varepsilon\right)} \qquad \text{for } \mathcal{L}^1\text{-a.e. } t \geq 0.$$

Thus, $\mu'_{u,\varepsilon}(t) \to \mu'_u(t)$ $\mathcal{L}^1$-a.e. in $[0,+\infty)$ as $\varepsilon \to 0$, since $|\nabla u|_{|\{u=t\}} \neq 0$ for $\mathcal{L}^1$-a.e. $t \geq 0$, in as much as $\dfrac{\mathcal{H}^{n-1}(\{u=t\})}{|\nabla u|_{|\{u=t\}}} \in L^1(0,\infty)$. This membership and the fact that

$$|\mu'_{u,\varepsilon}(t) - \mu'_u(t)| \leq \frac{\mathcal{H}^{n-1}(\{u=t\})}{|\nabla u|_{|\{u=t\}}} \qquad \text{for } \mathcal{L}^1\text{-a.e. } t \geq 0$$

entail that $\mu'_{u,\varepsilon} \to \mu'_u$ in $L^1(0,\infty)$. Hence, $\mu_{u,\varepsilon} \to \mu_u$ uniformly in $(0,\infty)$. Consequently, the functions $\mu_{u,\varepsilon} \circ u$ converge uniformly to $\mu_u \circ u$. Furthermore, by the chain rule for Sobolev functions (see e.g. [1, Theorem 3.96]),

$$\nabla(\mu_{u,\varepsilon} \circ u)(x) = \mu'_{u,\varepsilon}(u(x))\nabla u(x) = -\mathcal{H}^{n-1}(\{u=u(x)\})\frac{\nabla u(x)}{|\nabla u|_{|\{u=u(x)\}} + \varepsilon}$$

$$\text{for } \mathcal{L}^n\text{-a.e. } x \in U_0.$$

The last expression clearly converges to the right-hand side of (3.19). Moreover, from Theorem 3.1, we have that $\mathcal{H}^{n-1}(\{u = t\}) = \mathcal{H}^{n-1}(\partial U_t) \leq n\omega_n R_0^{n-1}$ for $\mathcal{L}^1$-a.e. $t \in (0, \operatorname{ess sup} u)$. Thus,

$$|\nabla(\mu_{u,\varepsilon} \circ u)(x)| \leq n\omega_n R_0^{n-1} \qquad \text{for } \mathcal{L}^n\text{-a.e. } x \in U_0.$$

By dominated convergence, $\nabla(\mu_{u,\varepsilon} \circ u)$ converges to the right-hand side of (3.19) in $L^1(U_0)$. Hence, the claim follows.

To conclude the proof let us now observe that for all $t \in (0, \operatorname{ess sup} u)$, $\mu_u(u(x)) \geq \mu_u(t) > 0$ for all $x \in U_0 \setminus \overline{U}_t$. Therefore, we can compute the derivatives of Φ in $U_0 \setminus \overline{U}_t$ by the usual chain rule formula for Sobolev functions, thus getting, from (3.19), that for $\mathcal{L}^n$-a.e. $x \in U_0 \setminus \overline{U}_t$

$$\nabla\Phi(x) = -\frac{1}{n\omega_n^{1/n}}(\mu_u(u(x)))^{\frac{1-n}{n}}\mathcal{H}^{n-1}(\{u=u(x)\})\frac{\nabla u(x)}{|\nabla u|_{|\{u=u(x)\}}}\chi_{\mathcal{D}_u^+}(x)$$

$$= -\frac{\nabla u(x)}{|\nabla u|_{|\{u=u(x)\}}}\chi_{\mathcal{D}_u^+}(x). \qquad (3.20)$$

Then, (3.18) follows immediately from Proposition 3.1 (ii). In particular, this proves that Φ is a $W^{1,\infty}$ function in the open set $U_0 \setminus \{u = \operatorname{ess sup} u\}$ (which is the difference of an open ball and a closed one). To conclude the proof it is enough to observe, that since $\mu_{u,\varepsilon} \circ u \in W^{1,\infty}(U_0)$, Φ has a continuous representative in U_0 which is constant on the closed ball $\{u = \text{esssup}\, u\}$. □

Notice that Lemma 3.3 is not telling us that Φ is Lipschitz continuous in U_0. It just says that Φ coincides $\mathcal{L}^n$-a.e. with a Lipschitz continuous function, with Lipschitz constant less than or equal to one. Therefore, we may only conclude that there exists a set N_0, with $\mathcal{L}^n(N_0) = 0$ such that

$$|\Phi(x) - \Phi(y)| \le |x - y| \qquad \text{for all } x, y \in U_0 \setminus N_0. \tag{3.21}$$

However, this information is enough to achieve the proof of Theorem 3.2.

Proof of Theorem 3.2. From the coarea formula (3.5), recalling (iv), (v) and (vi) of Proposition 3.1, we get

$$\int_0^{+\infty} \mathcal{H}^{n-1}(\partial U_t \cap N_0)\, dt = \int_{N_0} |\nabla u|\, dx = 0.$$

Therefore, there exists a set $I_0 \subset (0, +\infty)$, with $\mathcal{L}^1(I_0) = 0$, such that

$$\mathcal{H}^{n-1}(\partial U_t \cap N_0) = 0 \qquad \text{for all } t \in (0, +\infty) \setminus I_0.$$

Let us now fix $0 < s < t < \text{esssup}\, u$, with $s, t \notin I_0$. From Proposition 3.1 (v) we can find two sequences $\{x_h\} \subset \partial U_s \setminus N_0$ and $\{y_h\} \subset \partial U_t \setminus N_0$ such that

$$u(x_h) = s, \quad u(y_h) = t \quad \text{for all } h, \qquad |x_h - y_h| \to \text{dist}(\partial U_s, \partial U_t).$$

Since $\Phi(x_h) = R_s$ and $\Phi(y_h) = R_t$, from (3.21) we get that

$$|R_s - R_t| \le \lim_{h \to +\infty} |x_h - y_h| = \text{dist}(\partial U_s, \partial U_t)$$

Hence, U_s and U_t are concentric balls. From this one can easily conclude that U_s and U_t are indeed concentric for all $0 \le s < t \le \text{esssup}\, u$, thus proving the assertion. □

References

1. L.AMBROSIO, N.FUSCO & D.PALLARA, *Functions of bounded variation and free discontinuity problems,* Oxford University Press, Oxford, 2000
2. A. BAERNSTEIN II, *A unified approach to symmetrization,* in Partial differential equations of elliptic type, A. Alvino, E. Fabes & G. Talenti eds., Symposia Math. **35**, Cambridge Univ. Press, 1994
3. J.BOURGAIN, J.LINDENSTRAUSS & V.MILMAN, *Estimates related to Steiner symmetrizations,* in Geometric aspects of functional analysis, 264–273, Lecture Notes in Math. **1376**, Springer, Berlin, 1989
4. F.BROCK, *Weighted Dirichlet-type inequalities for Steiner Symmetrization,* Calc. Var., **8** (1999), 15–25

5. J.BROTHERS & W.ZIEMER, *Minimal rearrangements of Sobolev functions*, J. reine. angew. Math., **384** (1988), 153–179

6. YU.D.BURAGO & V.A.ZALGALLER, *Geometric inequalities*, Springer, Berlin, 1988

7. A.BURCHARD, *Steiner symmetrization is continuous in* $W^{1,p}$, Geom. Funct. Anal. **7** (1997), 823–860

8. E.CARLEN & M.LOSS, *Extremals of functionals with competing symmetries*, J. Funct. Anal. **88** (1990), 437–456

9. M.CHLEBIK, A.CIANCHI & N.FUSCO, *Perimeter inequalities for Steiner symmetrization: cases of equalities*, Annals of Math., **165** (2005), 525–555

10. A.CIANCHI & N.FUSCO, *Functions of bounded variation and rearrangements*, Arch. Rat. Mech. Anal. **165** (2002), 1–40

11. A.CIANCHI & N.FUSCO, *Minimal rearrangements, strict convexity and critical points*, to appear on Appl. Anal.

12. E.DE GIORGI, *Su una teoria generale della misura* $(r-1)$-*dimensionale in uno spazio a r dimensioni*, Ann. Mat. Pura Appl. (4), **36** (1954), 191–213

13. E.DE GIORGI, *Nuovi teoremi relativi alle misure* $(r-1)$-*dimensionali in uno spazio a r dimensioni*, Ricerche Mat., **4** (1955), 95–113

14. E.DE GIORGI, *Sulla proprietà isoperimetrica dell'ipersfera, nella classe degli insiemi aventi frontiera orientata di misura finita*, Atti Accad. Naz. Lincei Mem. Cl. Sci. Fis. Mat. Nat. Sez. I,**5** (1958), 33–44

15. L.C.EVANS & R.F.GARIEPY, *Lecture notes on measure theory and fine properties of functions*, CRC Press, Boca Raton, 1992

16. H.FEDERER, *Geometric measure theory*, Springer, Berlin, 1969

17. A.FERONE & R.VOLPICELLI, *Minimal rearrangements of Sobolev functions: a new proof*, Ann. Inst. H.Poincaré, Anal. Nonlinéaire **20** (2003), 333–339

18. M.GIAQUINTA, G.MODICA & J.SOUČEK, *Cartesian currents in the calculus of variations, Part I: Cartesian currents, Part II: Variational integrals*, Springer, Berlin, 1998

19. B.KAWOHL, *Rearrangements and level sets in PDE*, Lecture Notes in Math. **1150**, Springer, Berlin, 1985

20. B.KAWOHL, *On the isoperimetric nature of a rearrangement inequality and its consequences for some variational problems*, Arch. Rat. Mech. Anal. **94** (1986), 227–243

21. G.PÓLYA & G.SZEGÖ, *Isoperimetric inequalities in mathematical physics*, Annals of Mathematical Studies **27**, Princeton University Press, Princeton, 1951

22. STEINER, *Einfacher Beweis der isoperimetrischen Hauptsätze*, J. reine angew Math. 18 (1838), 281–296, and Gesammelte Werke, Vol. 2, 77–91, Reimer, Berlin, 1882

23. G.TALENTI, *Best constant in Sobolev inequality*, Ann. Mat. Pura Appl. **110** (1976), 353–372

24. G.TALENTI, *Nonlinear elliptic equations, rearrangements of functions and Orlicz spaces*, Ann. Mat. Pura Appl. **120** (1979), 159–184

25. G.TALENTI, *The standard isoperimetric theorem*, in Handbook of convex geometry, P.M.Gruber and J.M.Wills eds, North-Holland, Amsterdam, 1993

26. A.I.VOL'PERT, *Spaces BV and quasi-linear equations*, Math. USSR Sb., **17** (1967), 225–267

27. W.P.ZIEMER, *Weakly differentiable functions*, Springer, New York, 1989

CIME Courses on Partial Differential Equations and Calculus of Variations

Elvira Mascolo

Dipartimento di Matematica "U.Dini", Università di Firenze, viale Morgagni 67/A, 50134 Firenze, Italy
mascolo@math.unifi.it

David Hilbert used to say

> every real progress walks **hand in hand** with the discovery of more and more rigorous tools and simpler methods which meanwhile make easier the understanding of previous theories.

Nevertheless Augustus De Morgan used to say:

> The mental attitude which stimulate the mathematical invention is not only a sharp reasoning but rather a deep imagination.

The "progress" Hilbert was talking about is based – in mathematics more than in other scientific fields – upon teaching and collaboration and the "imagination" De Morgan was referring to, must be stimulated through a progressive and gradual learning.

Both history and everybody personal experience show that mathematical learning and its improvement is not just a matter of studying books and original articles, but rather that of a continuous and effective relationships with our own teacher(s) and collegues, rising new questions and discussing together their possible answers.

In the Fifties a group of outstanding Italian mathematicians, all member of the Scientific Committee of UMI (the Union of Italian Mathematicians) under the presidency of Enrico Bompiani, decided that it was the moment to rise the mathematical research in Italy to the level it was before the Second Worldly War and that it should be done through the organisation of high level courses. They realized the importance of providing the young researchers with the possibility of learning the new theories, subjects and themes which were appearing in those years and of mastering the new techniques and tools.

It was right in those years that the CIME was founded and the first course was held in Varenna (a charming small city on the Como lake) in 1954. The subject was on Functional Analysis, which can be considered at that time a new subject. More precisely:

Funzionali Analitici ed Anelli Normati
Varenna (Como), June 9–18, 1954

Lectures: L. Amerio (Politecnico Milano), L. Fantapié (Univ. Roma), E.R. Lorch (Columbia Univ.)

Seminars: M. Cugiani (Univ. Milano), F. Pellegrino (INAM, Roma), G.B. Rizza (Univ. Genova).

The second was also held in Varenna in the month of August of the same year. That was on

Quadratura Delle Superficie e Questioni Connesse
Varenna (Como), August 16–25, 1954

Lectures: R. Caccioppoli (Univ. Napoli), L. Cesari, (Univ. Bologna, Purdue Univ.), Chr.Y. Pauc (Univ. Rennes).

Seminars: A. Finzi (Technion, Haifa), A. Zygmund (Univ. Chicago).

The exceptional personalities were the teachers chosen in that occasion: Renato Caccioppoli, Lamberto Cesari and Antoni Zygmund. How can we recall without the suspect of being limited the contributions given for example by Caccioppoli to the development of the modern Functional Analysis and the Geometric Theory of Measure.

In the subsequent fifty-one years the CIME organized 163 courses covering basically every aspect of mathematics, both pure and applied, thus playing a crucial role in promoting and developing the mathematical research, not only in Italy. In fact the CIME activities have favoured and promoted personal contacts among distinguished scientists and young researchers, coming from all over the world.

The mathematics in last years has known a nearly explosive development and the organization of courses is an exceptional instrument of formation for the young investigators and a real support for the most mature ones.

The *full immersion*, permitted by a common location, is the right preamble to develop new subjects, to suggest new methods, to learn how to apply old methods to new problems, to start joint papers.

One main reason for the success of CIME courses was in particular the fact that they have been all published and the C.I.M.E. Sessions are an essential mean of diffusion of the mathematical culture.

The texts of lectures and seminars of each Session were all published:

- The volumes of Sessions 1-39 are actually out of print,
- The volumes of Sessions 39-70 are on the Catalogue of Edizioni Cremonese, Firenze, Italy
- The volumes of Sessions 71-83 are on the Catalogue of Liguori Editore, Napoli, Italy
- Since 1981 all courses notes are being published by Springer Verlag in a Subseries (Fondazione C.I.M.E.) of the Lectures Notes in Mathematics

My aim today is to guide you in a ideal journey through the development of the Calculus of Variation ad Nonlinear Differential Equation via the CIME courses held on these topics in the past fifty years of his history.

For the majority of younger people these arguments and techniques are to be considered as standard; however in these fifty years the developments have been so fast than the so-called "variational questions" raised by Hilbert at the beginning of the past century, have blowed up during the entire century (particularly the second half), in so many and different directions that Hilbert himself could have never imagined.

Quoting James Serrin we could say that

> the relevant field of investigation is nowadays so spread and wide that only a few years ago would have thought of as unbelivable...... more new ideas and results appeared from the end of the Second Worldly War until the present day than from the time of Talete until the year 1945.

The CIME courses on these subjects are all worthy of remark for the high scientific level of the directors and lectures.

In the following I will give a short presentation of each of them which allows to appreciate the significant role that the CIME Foundation played in the last 51 years.

- In 1958 Alessandro Faedo was the Scientific Direction of a session, strictly related with the variational questions and in particular devoted to the minima principles and their applications. v

Principio di Minimo e le sue Applicazioni in Analisi Funzionale
Pisa, September 1–10, 1958

Director: S. Faedo (Univ. Pisa).
Lectures: L. Bers (Courant Institute), Ch. B. Morrey Jr. (Univ. Of California, Berkeley), L. Nirenberg (Courant Institute).
Seminars: S. Agmon (Hebrew Univ. Jerusalem), G. Fichera (Univ. Roma), G. Stampacchia (Univ. Genova).
Notice the presence as lectures of Charles Jr. Morrey. and Louis Niremberg, two of the most important among the personalities in the Calculus of Variations and Partial Differential Equations of past century. The notes are published on Annali Scuola Normale di Pisa.

- In 1961 Enrico Bompiani was the Scientific Director of a session which can be considered a first approach to the geometric methods in the Calculus of Variations.

Geometria del Calcolo delle Variazioni
Saltino (Firenze), August 21–30, 1961

Director: E. Bompiani (Univ. Roma).

Lectures: H. Busemann (Univ. of Southern Calif., Los Angeles), E.T. Davies (Univ. Southampton), D. Laugwitz (Technische Hochschule, Darmstadt).

- In 1964, Guido Stampacchia was the Scientific Director of a very interesting session, in which we note the relevance not only the lecturers but also of the young researchers which gave a seminar.

Equazioni Differenziali non Lineari
Varenna (Como), August 30 – September 8, 1964

Director: G. Stampacchia (Univ. Pisa).
Lectures: P. Lax (New York Univ.), J. Leray (Collège De France), J. Moser (New York Univ.).
Seminars: R. Courant (New York Univ.), E. Degiorgi (Scuola Normale Superiore, Pisa), J. Friberg (Univ. Lund), J. Necas (CSAV, Praha), I. Segal (M.I.T.), G. Stampacchia (Univ. Pisa) O. Vejvoda (CSAV, Praha). Unfortunately there is not the text of the seminar given by Ennio de Giorgi.

- In 1966 Roberto Conti, one of the most important Italian mathematician and in some sense the *father* of CIME (he was Scientific Secretary since 1954 to 1974 and the Director of CIME Foundation since 1975 to 1998) was the Scientific Director of a session dedicated to the applications of the methods of Calculus of Variation to Control Theory that at that times was called "modern" Calculus of Variations.

Calculus of Variations, Classical and Modern
Bressanone (Bolzano), June 10–18, 1966

Director: R. Conti (Univ. Firenze).
Lectures: A. Blaquiere (Univ. Paris-Orsay), L. Cesari (Univ. Michigan), E. Rothe (Univ. Michigan), E. O. Roxin (Univ. Buenos Aires).
Seminars: C. Castaing (Univ. Caen), H. Halkin (Univ. California, La Jolla), C. Olech (Univ. Krakow).

- In 1972 Enrico Bombieri was the Scientific Director of a session dedicated to the Geometric measures. The lecturers are one of the most important in the field: Enrico Giusti, Frederik Almgren and Mario Miranda.

Geometric Measure Theory and Minimal Surfaces Varenna
(Como), August 25 – September 2, 1972

Director: E. Bombieri (Univ. Pisa).
Lectures: W.K. Allard (Princeton Univ.), F.J. Almgren (Princeton Univ.), E. Bombieri, E. Giusti (Univ. Pisa), M. Miranda (Univ. Ferrara).
Seminars: J. Guckenheimer (Princeton Univ.), D. Kinderlehrer (Univ. Minnesota), L. Piccinini (SNS, Pisa).

- In 1973 Guido Stampacchia and Gianfranco Capriz were the Scientific Directors of an important session on the Variational Methods in Mathematical Physics, which at that time were called "new".

New Variational Techniques in Mathematical Physics
Bressanone (Bolzano), June 17–26, 1973

Directors: G. Capriz (Univ. Pisa), G. Stampacchia (SNS, Pisa).
Lectures: G. Duvaut (Univ. Paris XIII), J.J. Moreau (Univ. Languedoc, Montpellier), B. Nayroles (Univ. Poitiers).
Seminars: C. Baiocchi (Univ. Pavia), Ch. Castaing (Univ. Languedoc, Montpellier), D. Kinderlehrer (Univ. Minnesota), H. Lanchon (Univ. Essex), J.M. Lasry (Univ. Paris-Dauphine), W. Noll (Carnegie Mellon Univ.), W. Velte (Univ. Wurzburg).

- In 1984 Enrico Giusti was the Scientific Director of a session devoted to the Harmonic Mapping.

Harmonic Mappings and Minimal Immersions
Montecatini Terme (Pistoia), June 24 – July 3, 1984

Director: E. Giusti (Univ. Firenze).
Lectures: S. Hildebrandt (Univ. Bonn), J. Jost (Univ. Bonn), L. Simon (Australian Nat. Univ., Canberra).
Seminars: J.H. Sampson (Johns Hopkins Univ.), M. Seppala (Univ. Helsinki).

- In 1987 Mariano Giaquinta was the Scientific Ddirector of a session in which different aspects of the Calculus of Variations were presented.

Topics in Calculus of Variations
Montecatini Terme (Pistoia), July 20–28, 1987

Director: M. Giaquinta (Univ. Firenze).
Lectures: L. Caffarelli (IAS, Princeton), A. J. Moser (ETH, Zurich), L. Nirenberg (Courant Inst.), R. Schoen (Univ. California, San Diego), A. Tromba (Max-Planck Inst., Bonn).
In this session for the first time Luis Caffarelli was a lecture. It is the begin of his valuable collaboration with C.I.M.E..

- In 1989 Arrigo Cellina was the Scientific Director of a session devoted to non convex problems and the methods for studying different subjects. In particular the non convex functionals of Calculus of Variations were considered.

Methods of Nonconvex Analysis
Varenna (Como), June 15–23, 1989

Director: A. Cellina (SISSA, Trieste).
Lectures: I. Ekeland (Univ. Paris, Dauphine), P. Marcellini (Univ. Firenze), A. Marino (Univ. Pisa), C. Olech (PAN, Warszawa), G. Pianigiani (Univ. Siena), T.R. Rockafellar (Univ. Washington, Seattle), M. Valadier (USTL, Montpellier).

- In 1995 Italo Capuzzo Dolcetta and Piere Louis Lions were the Scientific Directors of a session devoted to the viscosity solution and their applications in several fields of Partial Differential Equations.

Viscosity Solutions and Applications
Montecatini Terme (Pistoia), June 12–20, 1995

Directors: I. Capuzzo Dolcetta (Univ. Roma, La Sapienza), P.L. Lions (Univ. Paris Dauphine).
Lectures: M. Bardi (Univ. Padova), M.G. Crandall (Univ. California, Santa Barbara), L.C. Evans (Univ. California, Berkeley), M.H. Soner (Carnegie Mellon Univ.), P.E. Souganidis (Univ. Wisconsin).
One of the lecturers was L. C. Evans, which in the subsequent years has partecipated several times in the CIME activities.

- In 1996 Stefan Hildebrandt and Michael Struwe were Scientific Directors of a session, devoted to the contributions of the variational methods for the Ginzburg-Landau equations, the microstructure and phase transitions and the Plateau Problem.

Calculus of Variations and Geometric Evolution Problems
Cetraro (Cosenza), June 15–22, 1996

Directors: S. Hildebrandt (Univ. Bonn), M. Struwe (ETH, Zurich).
Lectures: F. Bethuel (ENS, Cachan), R. Hamilton (Univ. Califonia, San Diego), S. Muller (ETH, Zurich), K. Steffen (Univ. Dusseldorf).

- In 2001 Luis Caffarelli with Sandro Salsa were the Scientific Directors of a session which represents a basic guide on Optimal Transportation, by considering different point of views and perspectives,

Optimal Transportation and Applications
Martina Franca (Taranto), September 2–8, 2001

Directors: L. Caffarelli (Univ. Texas, Austin), S. Salsa (Politecnico Milano).
Lectures: L. Caffarelli (Univ. Texas, Austin), G. Buttazzo (Univ. Pisa), L.C. Evans (Univ. California, Berkeley), Y. Brenier (LAN-UPMC, Paris VI), C. Villani (Ecole Normale Sup., Lyon).

- Bernard Dacorogna and Paolo Marcellini are the Scientific Directors of this session, with almost 150 partecipants, the most attended course of the history of C.I.M.E.. Such wide partecipation is a clear indication that the Calculus of Variations is still an interesting and alive subject.

Calculus of Variations and NonLinear Partial Differential Equations
Cetraro (Cosenza), June 27 – July 2, 2005

Directors: Bernard Dacorogna (EPLF, Lausanne) Paolo Marcellini (Univ. Firenze).

Lectures L. Ambrosio (SNS Pisa), L.A. Caffarelli (Univ. Texas, Austin), M. Crandall (Univ. California, Santa Barbara), L.C. Evans (Univ. California, Berkeley, Usa), G. Dal Maso (SISSA, Trieste), N. Fusco (Univ. Napoli).

In these last years social and human sciences offered to the theory on Nonlinear PDE a new field to be added to the traditional ones coming from physics and natural sciences.

Even Hilbert in his famous conference in Paris in 1900, stressed the fruitfulness and opportunity of the interactions between *reason and experience* for the development of the mathematical theories.

However, nowadays it is a widespread opinion that pure mathematical research is important and absolutely necessary also without direct applications. Let me conclude by recalling the opinion of Cantor:

> The essence of mathematics is freedom and *independence*....
> freedom expressed as driving curiosity of a bright child....
> freedom to pursue innocent fascination until it finally touched the world we all live in.

List of Participants

1. Amar Micol
   Univ. Roma, La Sapienza, Italy
   `amar@dmmm.uniroma1.it`

2. Ambrosio Luigi
   SNS, Pisa, Berkeley, Italy
   `ambrosio@sns.it`
   (**lecturer**)

3. Angiuli Luciana
   Univ. Lecce, Italy
   `luciana.angiuli@unile.it`

4. Argiolas Roberto
   Univ. Cagliari, Italy
   `roberarg@unica.it`

5. Arsenio Diogo
   New York Univ., USA
   `arsenio@cims.nyu.edu`

6. Bandyopadhyay Saugata
   EPFL-Lausanne, Switzerland
   `saugata.bandyopadhyay@epfl.ch`

7. Barroso Ana Cristina
   Univ. Lisbon, Portugal
   `abarroso@ptmat.fc.ul.pt`

8. Benedetti Irene
   Univ. Firenze, Italy
   `benedetti@math.unifi.it`

9. Benincasa Elisabetta
   Univ. Calabria, Italy
   `benincasa@mat.unical.it`

10. Bertone Simone
    Univ. Milano-Bicocca, Italy
    `simone_bertone@libero.it`

11. Birindelli Isabeau
    Univ. Roma, La Sapienza, Italy
    `isabeau@mat.uniroma1.it`

12. Brancolini Alessio
    SNS, Pisa, Italy
    `a.brancolini@sns.it`

13. Briani Ariela
    Univ. Pisa, Italy
    `briani@dm.unipi.it`

14. Caffarelli Luis
    Univ. Texas at Austin, USA
    `caffarel@math.utexas.edu`
    (**lecturer**)

15. Cagnetti Filippo
    SISSA, Italy
    `cagnetti@sissa.it`

16. Castelpietra Marco
    Univ. Roma, Tor Vergata, Italy
    `castelpi@mat.uniroma2.it`

17. Celada Pietro
    Univ. Parma, Italy
    `pietro.celada@unipr.it`

18. Cellina Arrigo
    Univ. Milano Bicocca, Italy
    `cellina@matapp.unimib.it`

19. Cesaroni Annalisa
    Univ. Roma, La Sapienza, Italy
    `acesar@math.unipd.it`
20. Charro Caballero Fernando
    Univ. Autónoma Madrid, Spain
    `fernando.charro@uam.es`
21. Chermisi Milena
    Univ. Roma Tor Vergata, Italy
    `chermisi@mat.uniroma2.it`
22. Chinnici Marta
    Univ. Napoli, Italy
    `marta.chinnici@dma.unina.it`
23. Cicalese Marco
    Univ. Napoli, Italy
    `cicalese@unina.it`
24. Colucci Renato
    Univ. L'Aquila, Italy
    `coluccir@univaq.it`
25. Crandall Michael
    Univ. California, Santa Barbara
    `crandall@math.ucsb.edu`
    (**lecturer**)
26. Crippa Gianluca
    SNS, Pisa, Italy
    `g.crippa@sns.it`
27. Croce Gisella
    Univ. Montpellier II, France
    `croce@math.univ-montp2.fr`
28. Cupini Giovanni
    Univ. Firenze, Italy
    `cupini@math.unifi.it`
29. Dacorogna Bernard
    EPFL, Lausanne, Switzerland
    `Bernard.Dacorogna@epfl.ch`
    (**editor**)
30. Dal Maso Gianni
    SISSA, Italy
    `dalmaso@sissa.it`
31. Daryina Anna
    Russian Peoples Friendship
    University, Russia
    `daryina@ccas.ru`
32. Davini Andrea
    Univ. Pisa, Italy
    `davini@dm.unipi.it`

33. De Cicco Virginia
    Univ. Roma, La Sapienza, Italy
    `decicco@dmmm.uniroma1.it`
34. De Pascale Luigi
    Univ. Pisa, Italy
    `depascale@dm.unipi.it`
35. Defranceschi Anneliese
    Univ. Trento, Italy
    `defrance@science.unitn.it`
36. Della Pietra Francesco
    Univ. Napoli, Italy
    `dellapietra@dma.unina.it`
37. Demyanov Alexey
    St. Petersburg State University,
    USA
    `alex@ad9503.spb.edu`
38. Di Francesco Marco
    Univ. L'Aquila, Italy
    `difrance@univaq.it`
39. Di Nardo Rosaria
    Univ. Napoli, Federico II
    `rosaria.dinardo@dma.unina.it`
40. Donadello Carlotta
    SISSA-ISAS, Italy
    `donadel@ma.sissa.it`
41. Donatelli Donatella
    Univ. L'Aquila, Italy
    `donatell@univaq.it`
42. Donato Maria Bernadette
    Univ. Messina, Italy
    `bdonato@dipmat.unime.it`
43. Dragoni Federica
    SNS, Pisa, Italy
    `f.dragoni@sns.it`
44. Evans Craig L.
    Univ. California, Berkeley, USA
    `evans@math.berkeley.edu`
    (**lecturer**)
45. Ferriero Alessandro
    Ecole Polytechnique, France
    `ferriero@cmapx.polytechnique.fr`
46. Figalli Alessio
    SNS, Pisa, Italy
    `a.figalli@sns.it`

47. Focardi Matteo
    Univ. Firenze, Italy
    focardi@math.unifi.it
48. Fonte Massimo
    SISSA, Italy
    fonte@sissa.it
49. Forcadel Nicolas
    Cermics-ENPC, France
    forcadel@cermics.enpc.fr
50. Frosali Giovanni
    Univ. Firenze, Italy
    giovanni.frosali@unifi.it
51. Fusco Nicola
    Univ. Napoli, Italy
    n.fusco@unina.it
    (**lecturer**)
52. Garroni Adriana
    Univ. Roma, La Sapienza, Italy
    garroni@mat.uniroma1.it
53. Gavitone Nunzia
    Univ. Napoli, Italy
    nunzia.gavitone@dma.unina.it
54. Gebäck Tobias
    Chalmers Univ. of Technology
    Sweden
    tobiasg@math.chalmers.se
55. Gelli Maria Stella
    Univ. Pisa, Italy
    gelli@dm.unipi.it
56. Giorgieri Elena
    Univ. Roma, Tor Vergata, Italy
    giorgier@math.uniroma2.it
57. Granieri Luca
    Univ. Pisa, Italy
    granieri@mail.dm.unipi.it
58. Grasmair Markus
    Univ. Innsbruck, Austria
    markus.grasmair@uibk.ac.at
59. Grillo Gabriele
    Politecnico di Torino, Italy
    gabriele.grillo@polito.it
60. Grishchenko Alexey
    Voronezh State Univ., Russia
    alex_gr@rambler.ru
61. Iosifescu Oana
    Univ. Montpellier II, France
    iosifescu@math.univ-montp2.fr
62. Ivanov Denis
    St. Petersburg State Univ.,
    Russia
    denisiv@rol.ru
63. Kurzke Matthias
    IMA, Univ. Minnesota, USA
    kurzke@ima.umn.edu
64. Kuteeva Galina
    Saint-Petersburg State Univ.,
    Russia
    star@gk1662.spb.edu
65. La Russa Caterina
    Univ. Palermo, Italy
    larussa@math.unipa.it
66. Lattanzio Corrado
    Univ. L'Aquila, Italy
    corrado@univaq.it
67. Lebedyanskiy Victor
    Univ. Voronezh RIT, Russia
    leb_v@vpost.ru
68. Leonardi Gian Paolo
    Univ. Padova, Italy
    leonardi@dmsa.unipd.it
69. Leonetti Francesco
    Univ. L'Aquila, Italy
    leonetti@univaq.it
70. Leoni Fabiana
    Univ. Roma, La Sapienza, Italy
    leoni@mat.uniroma1.it
71. Leonori Tommaso
    Univ. Roma, Tor Vergata
    leonori@mat.uniroma2.it
72. Lewicka Marta
    Univ. Chicago, USA
    lewicka@math.uchicago.edu
73. Lorenz Thomas
    Univ. Heidelberg, Germany
    lorenz@iwr.uni-heidelberg.de
74. Loreti Paola
    Univ. Roma, La Sapienza, Italy
    loreti@dmmm.uniroma1.it

75. Lussardi Luca
Univ. Pavia, Italy
luca.lussardi@email.it

76. Lyakhova Sofya
Univ. Bristol, UK
s.lyakhova@bristol.ac.uk

77. Maddalena Francesco
Politecnico di Bari, Italy
maddalen@pascal.dm.uniba.it

78. Magnanini Rolando
Univ. Firenze, Italy
magnanin@math.unifi.it

79. Mannucci Paola
Univ. Padova, Italy
mannucci@math.unipd.it

80. Marcati Pierangelo
Univ. L'Aquila, Italy
marcati@univaq.it

81. Marcellini Paolo
Univ. Firenze, Italy
marcellini@math.unifi.it
(**editor**)

82. Marchini Elsa
CREA - École Polytechnique,
France
elsa.marchini@unimib.it

83. Marèchal Pierre
Univ. Toulouse 3, France
pierre.marechal@epfl.ch

84. Mariconda Carlo
Univ. Padova, Italy
maricond@math.unipd.it

85. Marigonda Antonio
Univ. Padova, Italy
amarigo@math.unipd.it

86. Mascolo Elvira
Univ. Firenze, Italy
mascolo@math.unifi.it

87. Matias José
Instituto Superior Tècnico,
Lisboa, Portugal
jmatias@math.ist.utl.pt

88. Mazzini Leonardo
Alenia Spazio, Italy
leonardo.mazzini@aleniaspazio.it

89. Milakis Emmanouil
Univ. Crete, Crete
milakis@math.uoc.gr

90. Milasi Monica
Univ. Messina, Italy
monica@dipmat.unime.it

91. Modica Giuseppe
Univ. Firenze, Italy
giuseppe.modica@unifi.it

92. Montanari Annamaria
Univ. Bologna, Italy
montanar@dm.unibo.it

93. Mora Maria Giovanna
SISSA, Italy
mora@sissa.it

94. Morini Massimiliano
SISSA, Italy
morini@sissa.it

95. Mugelli Francesco
Univ. Firenze, Italy
mugelli@math.unifi.it

96. Muratov Cyrill
NJIT, USA
muratov@njit.edu

97. Naumova Natalia
Saint-Petersburg State Univ.,
Russia
naumova@troll.phys.spbu.ru

98. Olech Michal
Univ. University Wroclaw
Poland
olech@math.uni.wroc.pl

99. Orlandi Giandomenico
Univ. Verona, Italy
orlandi@sci.univr.it

100. Panasenko Elena
Tambov State Univ., Russia
panlena_t@mail.ru

101. Papi Gloria
Univ. Firenze, Italy
papi@math.unifi.it

102. Paronetto Fabio
Univ. Lecce, Italy
fabio.paronetto@unile.it

103. Percivale Danilo
     Univ. Genova, Italy
     percivale@dimet.unige.it
104. Perrotta Stefania
     Univ. Modena e Reggio Emilia
     Italy
     perrotta@mail.unimo.it
105. Petitta Francesco
     Univ. Roma, La Sapienza, Italy
     petitta@mat.uniroma1.it
106. Pisante Adriano
     Univ. Roma, La Sapienza, Italy
     pisante@mat.uniroma1.it
107. Pisante Giovanni
     School of Maths, Georgia Tech.,
     USA
     pisante@math.gatech.edu
108. Poggiolini Laura
     Univ. Firenze, Italy
     poggiolini@math.unifi.it
109. Ponsiglione Marcello
     Max-Planck-Institute, Germany
     ponsigli@mis.mpg.de
110. Posilicano Andrea
     Univ. dell'Insubria, Italy
     posilicano@uninsubria.it
111. Pratelli Aldo
     Univ. Pavia, Italy
     aldo.pratelli@unipv.it
112. Primi Ivano
     Univ. Roma, La Sapienza, Italy
     primi@mat.uniroma1.it
113. Prinari Francesca
     Univ. Lecce, Italy
     prinari@mail.dm.unipi.it
114. Priuli Simone Fabio
     SISSA -ISAS, Italy
     priuli@sissa.it
115. Radice Teresa
     Univ. Napoli, Italy
     teresa.radice@dma.unina.it
116. Rampazzo Franco
     Univ. Padova, Italy
     rampazzo@math.unipd.it

117. Ribeiro Ana
     EPFL-Lausanne, Switzerland
     ana.ribeiro@epfl.ch
118. Riey Giuseppe
     Univ. Calabria, Italy
     riey@mat.unical.it
119. Roeger Matthias
     Technical Univ. Eindhoven,
     Nederlands
     mroeger@win.tue.nl
120. Rorro Marco
     Univ. Roma, La Sapienza, Italy
     rorro@caspur.it
121. Rossi Julio
     CSIC (Madrid, Spain)/
     Univ de Buenos Aires
     jrossi@dm.uba.ar
122. Rubino Bruno
     Univ. L'Aquila, Italy
     rubino@ing.univaq.it
123. Santambrogio Filippo
     SNS, Pisa, Italy
     santambrogio@sns.it
124. Scardia Lucia
     SISSA, Italy
     scardia@sissa.it
125. Segatti Antonio
     Univ. Pavia, Italy
     antonio.segatti@unipv.it
126. Serapioni Raul
     Univ. Trento, Italy
     serapion@science.unitn.it
127. Shishkina Elina
     Voronezh St. Tech Academy,
     Russia
     ilina_dico@mail.ru
128. Siconolfi Antonio
     Univ. Roma, La Sapienza, Italy
     siconolf@mat.uniroma1.it
129. Silvestre Luis
     Univ. Texas at Austin, USA
     lsilvest@math.utexas.edu
130. Sirakov Boyan
     Univ. Paris X, Italy
     sirakov@ehess.fr

131. Solferino Viviana
    Univ. Calabria, Italy
    solferino@mat.unical.it

132. Soravia Pierpaolo
    Univ. Padova, Italy
    soravia@math.unipd.it

133. Spinolo Laura Valentina
    SISSA, Italy
    spinolo@sissa.it

134. Stefanelli Ulisse
    CNR, Italy
    ulisse@imati.cnr.it

135. Stroffolini Bianca
    Univ. Napoli, Italy
    bstroffo@unina.it

136. Terra Joana
    Univ. Politecnica de Catalunya,
    Spain
    joana.terra@upc.edu

137. Terrone Gabriele
    Univ. Padova, Italy
    gabter@math.unipd.it

138. Treu Giulia
    Univ. Padova, Italy
    treu@math.unipd.it

139. Turbin Mikhail
    Voronezh State Univ., Russia
    mrmike@math.vsu.ru

140. Veneroni Marco
    Univ. Pavia, Italy
    marco.veneroni@unipv.it

141. Verzini Gianmaria
    Politecnico di Milano, Italy
    gianmaria.verzini@polimi.it

142. Villa Elena
    Univ. Milano, Italy
    villa@mat.unimi.it

143. Villa Silvia
    Univ. Firenze, Italy
    villa@math.unifi.it

144. Vitali Enrico
    Univ. Pavia, Italy
    enrico.vitali@unipv.it

145. Zanini Chiara
    SISSA-ISAS, Italy
    zaninic@sissa.it

146. Zappale Elvira
    Univ. Salerno, Italy
    zappale@bridge.diima.unisa.it

147. Zeppieri Caterina
    Univ. Roma, La Sapienza, Italy
    zeppieri@mat.uniroma1.it

LIST OF C.I.M.E. SEMINARS

Published by C.I.M.E

Published by Ed. Cremonese, Firenze

1966   39. Calculus of variations
          40. Economia matematica
          41. Classi caratteristiche e questioni connesse
          42. Some aspects of diffusion theory

1967   43. Modern questions of celestial mechanics
          44. Numerical analysis of partial differential equations
          45. Geometry of homogeneous bounded domains

1968   46. Controllability and observability
          47. Pseudo-differential operators
          48. Aspects of mathematical logic

1969   49. Potential theory
          50. Non-linear continuum theories in mechanics and physics and their applications
          51. Questions of algebraic varieties

1970   52. Relativistic fluid dynamics
          53. Theory of group representations and Fourier analysis
          54. Functional equations and inequalities
          55. Problems in non-linear analysis

1971   56. Stereodynamics
          57. Constructive aspects of functional analysis (2 vol.)
          58. Categories and commutative algebra

1972   59. Non-linear mechanics
          60. Finite geometric structures and their applications
          61. Geometric measure theory and minimal surfaces

1973   62. Complex analysis
          63. New variational techniques in mathematical physics
          64. Spectral analysis

1974   65. Stability problems
          66. Singularities of analytic spaces
          67. Eigenvalues of non linear problems

1975   68. Theoretical computer sciences
          69. Model theory and applications
          70. Differential operators and manifolds

Published by Ed. Liguori, Napoli

1976   71. Statistical Mechanics
          72. Hyperbolicity
          73. Differential topology

1977   74. Materials with memory
          75. Pseudodifferential operators with applications
          76. Algebraic surfaces

Published by Ed. Liguori, Napoli & Birkhäuser

1978   77. Stochastic differential equations
          78. Dynamical systems

1979   79. Recursion theory and computational complexity
          80. Mathematics of biology

Published by Springer-Verlag

Lecture Notes in Mathematics

For information about earlier volumes
please contact your bookseller or Springer
LNM Online archive: springerlink.com

Vol. 1835: O.T. Izhboldin, B. Kahn, N.A. Karpenko, A. Vishik, Geometric Methods in the Algebraic Theory of Quadratic Forms. Summer School, Lens, 2000. Editor: J.-P. Tignol (2004)

Vol. 1836: C. Năstăsescu, F. Van Oystaeyen, Methods of Graded Rings. XIII, 304 p, 2004.

Vol. 1837: S. Tavaré, O. Zeitouni, Lectures on Probability Theory and Statistics. Ecole d'Eté de Probabilités de Saint-Flour XXXI-2001. Editor: J. Picard (2004)

Vol. 1838: A.J. Ganesh, N.W. O'Connell, D.J. Wischik, Big Queues. XII, 254 p, 2004.

Vol. 1839: R. Gohm, Noncommutative Stationary Processes. VIII, 170 p, 2004.

Vol. 1840: B. Tsirelson, W. Werner, Lectures on Probability Theory and Statistics. Ecole d'Eté de Probabilités de Saint-Flour XXXII-2002. Editor: J. Picard (2004)

Vol. 1841: W. Reichel, Uniqueness Theorems for Variational Problems by the Method of Transformation Groups (2004)

Vol. 1842: T. Johnsen, A. L. Knutsen, K_3 Projective Models in Scrolls (2004)

Vol. 1843: B. Jefferies, Spectral Properties of Noncommuting Operators (2004)

Vol. 1844: K.F. Siburg, The Principle of Least Action in Geometry and Dynamics (2004)

Vol. 1845: Min Ho Lee, Mixed Automorphic Forms, Torus Bundles, and Jacobi Forms (2004)

Vol. 1846: H. Ammari, H. Kang, Reconstruction of Small Inhomogeneities from Boundary Measurements (2004)

Vol. 1847: T.R. Bielecki, T. Björk, M. Jeanblanc, M. Rutkowski, J.A. Scheinkman, W. Xiong, Paris-Princeton Lectures on Mathematical Finance 2003 (2004)

Vol. 1848: M. Abate, J. E. Fornaess, X. Huang, J. P. Rosay, A. Tumanov, Real Methods in Complex and CR Geometry, Martina Franca, Italy 2002. Editors: D. Zaitsev, G. Zampieri (2004)

Vol. 1849: Martin L. Brown, Heegner Modules and Elliptic Curves (2004)

Vol. 1850: V. D. Milman, G. Schechtman (Eds.), Geometric Aspects of Functional Analysis. Israel Seminar 2002-2003 (2004)

Vol. 1851: O. Catoni, Statistical Learning Theory and Stochastic Optimization (2004)

Vol. 1852: A.S. Kechris, B.D. Miller, Topics in Orbit Equivalence (2004)

Vol. 1853: Ch. Favre, M. Jonsson, The Valuative Tree (2004)

Vol. 1854: O. Saeki, Topology of Singular Fibers of Differential Maps (2004)

Vol. 1855: G. Da Prato, P.C. Kunstmann, I. Lasiecka, A. Lunardi, R. Schnaubelt, L. Weis, Functional Analytic Methods for Evolution Equations. Editors: M. Iannelli, R. Nagel, S. Piazzera (2004)

Vol. 1856: K. Back, T.R. Bielecki, C. Hipp, S. Peng, W. Schachermayer, Stochastic Methods in Finance, Bressanone/Brixen, Italy, 2003. Editors: M. Fritelli, W. Runggaldier (2004)

Vol. 1857: M. Émery, M. Ledoux, M. Yor (Eds.), Séminaire de Probabilités XXXVIII (2005)

Vol. 1858: A.S. Cherny, H.-J. Engelbert, Singular Stochastic Differential Equations (2005)

Vol. 1859: E. Letellier, Fourier Transforms of Invariant Functions on Finite Reductive Lie Algebras (2005)

Vol. 1860: A. Borisyuk, G.B. Ermentrout, A. Friedman, D. Terman, Tutorials in Mathematical Biosciences I. Mathematical Neurosciences (2005)

Vol. 1861: G. Benettin, J. Henrard, S. Kuksin, Hamiltonian Dynamics – Theory and Applications, Cetraro, Italy, 1999. Editor: A. Giorgilli (2005)

Vol. 1862: B. Helffer, F. Nier, Hypoelliptic Estimates and Spectral Theory for Fokker-Planck Operators and Witten Laplacians (2005)

Vol. 1863: H. Führ, Abstract Harmonic Analysis of Continuous Wavelet Transforms (2005)

Vol. 1864: K. Efstathiou, Metamorphoses of Hamiltonian Systems with Symmetries (2005)

Vol. 1865: D. Applebaum, B.V. R. Bhat, J. Kustermans, J. M. Lindsay, Quantum Independent Increment Processes I. From Classical Probability to Quantum Stochastic Calculus. Editors: M. Schürmann, U. Franz (2005)

Vol. 1866: O.E. Barndorff-Nielsen, U. Franz, R. Gohm, B. Kümmerer, S. Thorbjønsen, Quantum Independent Increment Processes II. Structure of Quantum Lévy Processes, Classical Probability, and Physics. Editors: M. Schürmann, U. Franz, (2005)

Vol. 1867: J. Sneyd (Ed.), Tutorials in Mathematical Biosciences II. Mathematical Modeling of Calcium Dynamics and Signal Transduction. (2005)

Vol. 1868: J. Jorgenson, S. Lang, $Pos_n(R)$ and Eisenstein Series. (2005)

Vol. 1869: A. Dembo, T. Funaki, Lectures on Probability Theory and Statistics. Ecole d'Eté de Probabilités de Saint-Flour XXXIII-2003. Editor: J. Picard (2005)

Vol. 1870: V.I. Gurariy, W. Lusky, Geometry of Müntz Spaces and Related Questions. (2005)

Vol. 1871: P. Constantin, G. Gallavotti, A.V. Kazhikhov, Y. Meyer, S. Ukai, Mathematical Foundation of Turbulent Viscous Flows, Martina Franca, Italy, 2003. Editors: M. Cannone, T. Miyakawa (2006)

Vol. 1872: A. Friedman (Ed.), Tutorials in Mathematical Biosciences III. Cell Cycle, Proliferation, and Cancer (2006)

Vol. 1873: R. Mansuy, M. Yor, Random Times and Enlargements of Filtrations in a Brownian Setting (2006)

Vol. 1874: M. Yor, M. Émery (Eds.), In Memoriam Paul-André Meyer - Séminaire de Probabilités XXXIX (2006)

Vol. 1875: J. Pitman, Combinatorial Stochastic Processes. Ecole d'Eté de Probabilités de Saint-Flour XXXII-2002. Editor: J. Picard (2006)

Vol. 1876: H. Herrlich, Axiom of Choice (2006)

Vol. 1877: J. Steuding, Value Distributions of L-Functions (2007)

Vol. 1878: R. Cerf, The Wulff Crystal in Ising and Percolation Models, Ecole d'Eté de Probabilités de Saint-Flour XXXIV-2004. Editor: Jean Picard (2006)

Vol. 1879: G. Slade, The Lace Expansion and its Applications, Ecole d'Eté de Probabilités de Saint-Flour XXXIV-2004. Editor: Jean Picard (2006)

Vol. 1880: S. Attal, A. Joye, C.-A. Pillet, Open Quantum Systems I, The Hamiltonian Approach (2006)

Vol. 1881: S. Attal, A. Joye, C.-A. Pillet, Open Quantum Systems II, The Markovian Approach (2006)

Vol. 1882: S. Attal, A. Joye, C.-A. Pillet, Open Quantum Systems III, Recent Developments (2006)

Vol. 1883: W. Van Assche, F. Marcellàn (Eds.), Orthogonal Polynomials and Special Functions, Computation and Application (2006)

Vol. 1884: N. Hayashi, E.I. Kaikina, P.I. Naumkin, I.A. Shishmarev, Asymptotics for Dissipative Nonlinear Equations (2006)

Vol. 1885: A. Telcs, The Art of Random Walks (2006)

Vol. 1886: S. Takamura, Splitting Deformations of Degenerations of Complex Curves (2006)

Vol. 1887: K. Habermann, L. Habermann, Introduction to Symplectic Dirac Operators (2006)

Vol. 1888: J. van der Hoeven, Transseries and Real Differential Algebra (2006)

Vol. 1889: G. Osipenko, Dynamical Systems, Graphs, and Algorithms (2006)

Vol. 1890: M. Bunge, J. Funk, Singular Coverings of Toposes (2006)

Vol. 1891: J.B. Friedlander, D.R. Heath-Brown, H. Iwaniec, J. Kaczorowski, Analytic Number Theory, Cetraro, Italy, 2002. Editors: A. Perelli, C. Viola (2006)

Vol. 1892: A. Baddeley, I. Bárány, R. Schneider, W. Weil, Stochastic Geometry, Martina Franca, Italy, 2004. Editor: W. Weil (2007)

Vol. 1893: H. Hanßmann, Local and Semi-Local Bifurcations in Hamiltonian Dynamical Systems, Results and Examples (2007)

Vol. 1894: C.W. Groetsch, Stable Approximate Evaluation of Unbounded Operators (2007)

Vol. 1895: L. Molnár, Selected Preserver Problems on Algebraic Structures of Linear Operators and on Function Spaces (2007)

Vol. 1896: P. Massart, Concentration Inequalities and Model Selection, Ecole d'Été de Probabilités de Saint-Flour XXXIII-2003. Editor: J. Picard (2007)

Vol. 1897: R. Doney, Fluctuation Theory for Lévy Processes, Ecole d'Été de Probabilités de Saint-Flour XXXV-2005. Editor: J. Picard (2007)

Vol. 1898: H.R. Beyer, Beyond Partial Differential Equations, On linear and Quasi-Linear Abstract Hyperbolic Evolution Equations (2007)

Vol. 1899: Séminaire de Probabilités XL. Editors: C. Donati-Martin, M. Émery, A. Rouault, C. Stricker (2007)

Vol. 1900: E. Bolthausen, A. Bovier (Eds.), Spin Glasses (2007)

Vol. 1901: O. Wittenberg, Intersections de deux quadriques et pinceaux de courbes de genre 1, Intersections of Two Quadrics and Pencils of Curves of Genus 1 (2007)

Vol. 1902: A. Isaev, Lectures on the Automorphism Groups of Kobayashi-Hyperbolic Manifolds (2007)

Vol. 1903: G. Kresin, V. Maz'ya, Sharp Real-Part Theorems (2007)

Vol. 1904: P. Giesl, Construction of Global Lyapunov Functions Using Radial Basis Functions (2007)

Vol. 1905: C. Prévôt, M. Röckner, A Concise Course on Stochastic Partial Differential Equations (2007)

Vol. 1906: T. Schuster, The Method of Approximate Inverse: Theory and Applications (2007)

Vol. 1907: M. Rasmussen, Attractivity and Bifurcation for Nonautonomous Dynamical Systems (2007)

Vol. 1908: T.J. Lyons, M. Caruana, T. Lévy, Differential Equations Driven by Rough Paths, Ecole d'Été de Probabilités de Saint-Flour XXXIV-2004 (2007)

Vol. 1909: H. Akiyoshi, M. Sakuma, M. Wada, Y. Yamashita, Punctured Torus Groups and 2-Bridge Knot Groups (I) (2007)

Vol. 1910: V.D. Milman, G. Schechtman (Eds.), Geometric Aspects of Functional Analysis. Israel Seminar 2004-2005 (2007)

Vol. 1911: A. Bressan, D. Serre, M. Williams, K. Zumbrun, Hyperbolic Systems of Balance Laws. Lectures given at the C.I.M.E. Summer School held in Cetraro, Italy, July 14–21, 2003. Editor: P. Marcati (2007)

Vol. 1912: V. Berinde, Iterative Approximation of Fixed Points (2007)

Vol. 1913: J.E. Marsden, G. Misiołek, J.-P. Ortega, M. Perlmutter, T.S. Ratiu, Hamiltonian Reduction by Stages (2007)

Vol. 1914: G. Kutyniok, Affine Density in Wavelet Analysis (2007)

Vol. 1915: T. Bıyıkoğlu, J. Leydold, P.F. Stadler, Laplacian Eigenvectors of Graphs. Perron-Frobenius and Faber-Krahn Type Theorems (2007)

Vol. 1916: C. Villani, F. Rezakhanlou, Entropy Methods for the Boltzmann Equation. Editors: F. Golse, S. Olla (2008)

Vol. 1917: I. Veselić, Existence and Regularity Properties of the Integrated Density of States of Random Schrödinger (2008)

Vol. 1918: B. Roberts, R. Schmidt, Local Newforms for GSp(4) (2007)

Vol. 1919: R.A. Carmona, I. Ekeland, A. Kohatsu-Higa, J.-M. Lasry, P.-L. Lions, H. Pham, E. Taflin, Paris-Princeton Lectures on Mathematical Finance 2004. Editors: R.A. Carmona, E. Çinlar, I. Ekeland, E. Jouini, J.A. Scheinkman, N. Touzi (2007)

Vol. 1920: S.N. Evans, Probability and Real Trees. Ecole d'Été de Probabilités de Saint-Flour XXXV-2005 (2008)

Vol. 1921: J.P. Tian, Evolution Algebras and their Applications (2008)

Vol. 1922: A. Friedman (Ed.), Tutorials in Mathematical BioSciences IV. Evolution and Ecology (2008)

Vol. 1923: J.P.N. Bishwal, Parameter Estimation in Stochastic Differential Equations (2008)

Vol. 1924: M. Wilson, Littlewood-Paley Theory and Exponential-Square Integrability (2008)

Vol. 1925: M. du Sautoy, L. Woodward, Zeta Functions of Groups and Rings (2008)

Vol. 1926: L. Barreira, V. Claudia, Stability of Nonautonomous Differential Equations (2008)

Vol. 1927: L. Ambrosio, L. Caffarelli, M.G. Crandall, L.C. Evans, N. Fusco, Calculus of Variations and Non-Linear Partial Differential Equations. Lectures given at the C.I.M.E. Summer School held in Cetraro, Italy, June 27–July 2, 2005. Editors: B. Dacorogna, P. Marcellini (2008)

Vol. 1928: J. Jonsson, Simplicial Complexes of Graphs (2008)

Recent Reprints and New Editions

Vol. 1618: G. Pisier, Similarity Problems and Completely Bounded Maps. 1995 – 2nd exp. edition (2001)

Vol. 1629: J.D. Moore, Lectures on Seiberg-Witten Invariants. 1997 – 2nd edition (2001)

Vol. 1638: P. Vanhaecke, Integrable Systems in the realm of Algebraic Geometry. 1996 – 2nd edition (2001)

Vol. 1702: J. Ma, J. Yong, Forward-Backward Stochastic Differential Equations and their Applications. 1999 – Corr. 3rd printing (2007)

Vol. 830: J.A. Green, Polynomial Representations of GL_n, with an Appendix on Schensted Correspondence and Littelmann Paths by K. Erdmann, J.A. Green and M. Schocker 1980 – 2nd corr. and augmented edition (2007)

Printed in the United States
By Bookmasters